Test Bank to Accompany

# CONCEPTUAL
# Physics

## FIFTH EDITION

**Test Bank to Accompany**

# CONCEPTUAL
# Physics

## FIFTH EDITION

City College of San Francisco

Little, Brown and Company
Boston   Toronto

ISBN 0-316-35976-9


5  4  3  2  1

ALP

Published simultaneously in Canada
by Little, Brown & Company (Canada) Limited

Printed in the United States of America

MAKETEST is a trademark of Annette Rappleyea.

IBM is a registered trademark of International Business Machines Corporation.

Apple is a registered trademark of Apple Computer Inc.

Honeywell is a registered trademark of Honeywell, Inc.

# Acknowledgments

Much appreciation and thanks goes to Annette Rappleyea, a friend and colleague of mine at CCSF, for coordinating this edition of the test bank. She spent many hours typing questions, editing questions, and adding questions of her own.

Thanks also goes to John Hubisz of College of the Mainland in Texas City, Texas, for editing questions and making many helpful suggestions.

Finally I thank two other CCSF colleagues of mine; Dack Lee, for carefully checking over the questions, and Betty Mattea for the use of her printer.

# Contents

A.  Introduction ........................................................ 1

B.  Preparing the Computer and Your Diskettes ...................... 3

    1. Starting the Computer ....................................... 3

    2. Protecting Your Diskettes ................................... 3

        a. Formatting a Diskette ................................... 4

        b. Copying a Diskette ...................................... 4

    3. Seeing What's on a Diskette ................................. 5

    4. Loading a Program ........................................... 5

    5. Listing a Chapter File ...................................... 6

    6. Seeing How Much Space is on a Diskette ...................... 7

C.  Running MAKETEST ............................................... 8

    1. Starting MAKETEST ........................................... 8

    2. The MAKETEST Menu ........................................... 8

    3. Creating a Test ............................................. 9

        a. Choosing Specific Questions ............................. 9

            i.  Choosing the Chapters ............................... 9

            11. Choosing the Questions ............................. 10

        b. Letting the Computer Choose Questions at Random ........ 10

            i.  Choosing the Chapters ............................... 10

            11. Choosing the Questions ............................. 11

        c. Escape from Create Test Option ......................... 11

    4. Viewing a Test .............................................. 11

    5. Printing a Test ............................................. 12

    6. Reordering a Test ........................................... 13

    7. Ending the Session .......................................... 13

D.  Running EDITBANK ............................................... 14

    1. Starting EDITBANK ........................................... 14

    2. The EDITBANK Menu ........................................... 14

    3. Creating a New File ......................................... 15

        a. Entering the Chapter Number ............................ 15

      b. Entering the Questions ................................... 15

      c. Entering the Answer ..................................... 16

      d. Entering the Difficulty Level .......................... 16

      e. Closing the File ....................................... 17

   4. Adding to a File .......................................... 17

      a. Entering the Questions ................................. 17

      b. Entering the Answer .................................... 17

      c. Entering the Difficulty Level .......................... 18

      d. Closing the File ....................................... 18

   5. Editing a File ............................................ 18

      a. Choosing to Edit a Question ............................ 18

      b. Editing a Question ..................................... 19

      c. Changing the Answer .................................... 19

      d. Changing the Difficulty Level .......................... 19

      e. Closing the File ....................................... 20

   6. Updating a File ........................................... 20

      a. Using a Different Editor ............................... 20

      b. Using the Update Option ................................ 21

   7. Ending the Session ........................................ 21

E. Appendices ................................................... 22

   1. Appendix A (question code) ................................ 22

      a. The First Numbers ...................................... 23

      b. The Code ............................................... 24

      c. The Rest of the File ................................... 24

   2. Appendix B (printer codes) ................................ 24

F. The Test Bank ................................................ 26

Summary of Questions ............................................ 27

Part 1: Mechanics

   Chapter 2: Linear Motion ..................................... 29

   Chapter 3a: Projectiles ...................................... 34

   Chapter 3b: Satellite Motion ................................. 36

   Chapter 4a: Non-Free Fall .................................... 39

Chapter 4b: Newton's First and Second Laws of Motion ....... 42

Chapter 4c: Newton's Third Law of Motion ................... 49

Chapter 5: Momentum ....................................... 52

Chapter 6a: Work and Energy ............................... 57

Chapter 6b: Momentum and Energy ........................... 62

Chapter 7: Rotational Motion .............................. 65

Chapter 8a: Gravitation ................................... 69

Chapter 8b: Tides ......................................... 72

Part 2: Properties of Matter.

Chapter 9: The Atomic Nature of Matter .................... 74

Chapter 10a: Solids ....................................... 77

Chapter 10b: Scaling ...................................... 80

Chapter 11a: Liquids ...................................... 83

Chapter 11b: Pascal's Principle and Surface Tension ....... 89

Chapter 12a: Gases ........................................ 91

Chapter 12b: Plasmas ...................................... 95

Part 3: Heat

Chapter 13: Temperature, Heat and Expansion ............... 96

Chapter 14: Heat Transfer ................................. 99

Chapter 15: Change of State ............................... 101

Chapter 16: Thermodynamics ................................ 104

Part 4: Sound

Chapter 17: Vibrations and Waves .......................... 108

Chapter 18: Sound ......................................... 112

Chapter 19: Music ......................................... 115

Part 5: Electricity and Magnetism

Chapter 20: Electrostatics ................................ 117

Chapter 21: Electric Current .............................. 121

Chapter 22: Magnetism ..................................... 125

Chapter 23: Electromagnetic Interactions ................. 127

Chapter 24: Electromagnetic Radiation .................... 129

Part 6: Light

    Chapter 25: Reflection and Refraction ..................... 132

    Chapter 26: Color ........................................ 137

    Chapter 27: Interference and Polarization ................ 140

    Chapter 28: Light Emission ............................... 142

    Chapter 29: Light Quanta ................................. 145

Part 7: Atomic and Nuclear Physics

    Chapter 30: Atoms and Quanta ............................. 147

    Chapter 31: The Atomic Nucleus and Radioactivity ......... 151

    Chapter 32: Nuclear Fission and Fusion ................... 155

Part 8: Relativity and Astrophysics

    Chapter 33: The Special Theory of Relativity ............. 159

    Chapter 34: The General Theory of Relativity ............. 163

    Chapter 35: Astrophysics ................................. 166

Appendix

    Appendix 5: Exponential Growth ........................... 170

# MAKETEST™

(Test Making Program)

Annette Rappleyea
City College of San Francisco

# Introduction

Welcome to the CONCEPTUAL PHYSICS test bank and the test-making programs, MAKETEST and EDITBANK! The test bank has been created by Paul Hewitt at City College of San Francisco. The test-making programs have been created specifically for use with the test bank and have been written by me, Annette Rappleyea, a colleague and friend of Paul's at CCSF.

Many hours have been spent on the test bank questions. There are over 1000 questions. Most of them are Paul's - ingeniously written in his own style. Some of these questions have come from me and a few have come from John Hubisz, a professor at College of the Mainland in Texas City, Texas.

Even though the test bank covers most of the concepts in Paul's book, there are still lots of places where extra questions can be added. Since every teacher will have favorite questions he or she will want to include on a test, each chapter's test bank can be modified by using the program, EDITBANK. In addition, there are six "empty" chapters that do not correspond to chapters in the book that can be filled with new questions.

Questions are grouped by order of difficulty, A, B, or C. Type A questions are the easiest and are taken directly from easy-to-find references in the text. Type B questions are more difficult, but can also be answered by a careful reading of the text. Type C questions are the most difficult, and require more thought than either of the other two types.

Questions are also grouped according to whether or not there is a numerical calulation involved in them. Those involving a calculation are labeled with an "M" in the test bank booklet.

Answers in the test-bank booklet are shown in parentheses and the difficulty level is shown to the right of the answer. For example a problem with these letters after it,

(b)CM

would be a C level problem that includes a calculation and has an answer, b.

MAKETEST was first written several years ago to help Paul in his conceptual physics classes. Each semester Paul laboriously composed problems on a typewriter, then cut and pasted problems to make tests. Besides being a lot of work, it took a lot of time, especially since Paul usually has very large classes and uses many tests. After convincing him that it would be a lot faster using a computer, I wrote the first version of MAKETEST. It ran well and for several semesters Paul used it to produce many tests. At that time MAKETEST ran on a mainframe Honeywell® computer. Turn-around time was slow and test-bank questions took up valuable disk space. With the arrival of microcomputers at City College, we started thinking about putting the program on an IBM® personal computer. At the same time, Ron Pullins, the science Editor at Little, Brown and Co, became interested in distributing a test generator as a teaching aid to go with the fifth edition of Paul's book.

The main goal of MAKETEST and EDITBANK is to make testmaking easier for the instructor and to reduce the amount of time it takes to make a test.

We have included complete instructions for use of MAKETEST and EDITBANK in this booklet. These instructions are written for running the programs on an IBM personal computer. At the time of this printing, both programs are being rewritten for an Apple® II computer. Additional documentation for the Apple will be available on request from Little, Brown and Company.

The test bank, MAKETEST and EDITBANK have been copyrighted and licensed for distribution to Little, Brown and Company. If you want a copy of the programs or the test bank, write:

> Elizabeth Phillips, Sales Administrator, or
> Ron Pullins, Science Editor
> Little, Brown and Company
> 34 Beacon Street
> Boston, Massachusetts 02106

Customized testing is also available for adopters without access to compatible microcomputers. For further details please inquire to:

> Wayne Strickland, Product Manager
> Little, Brown and Company
> 34 Beacon Street
> Boston, Massachusetts 02106

If you have questions or suggestions about questions in the test bank contact:

> Paul Hewitt, Box S-60
> City College of San Francisco
> 50 Phelan Avenue
> San Francisco, California 94112
> (415-239-3375).

If you have technical questions on running either MAKETEST or EDITBANK, contact me:

> Annette Rappleyea, Box S-120
> City College of San Francisco
> 50 Phelan Avenue
> San Francisco, California 94112
> (415-239-3473 or 415-239-3621).

Have fun using these programs!

MAKETEST software is available only to adopters who have ordered texts from Little, Brown and Company.

# Preparing the Computer and Your Diskettes

Before running MAKETEST or EDITBANK, you need to set up the computer to be able to handle the program, and also you need to take some precautions with the diskettes.  How to do these things is outlined in the following sections.

### 1.  STARTING THE COMPUTER

The first thing to do is to put a copy of your operating system into drive a: of the computer and turn on the power.  Both MAKETEST and EDITBANK will run with Microsoft DOS 1.1 or DOS 2.0.  Once the operating system is running you will see a prompt that looks like this:

A>

This means the operating system has been loaded and is ready to do the busy work involved in running a program (This procedure is called "booting up" the machine.).

The default drive is the drive that appears in the DOS prompt, A>.  In most cases it will be drive a:.

If your computer is already on, you can re-boot it by putting the DOS diskette in drive a: and pushing the "Alt" key (lower left side of keyboard), the "Ctrl" key (middle left side of keyboard) and the "Del" key (below number pad) all at the same time.

In the process of booting up, you will be prompted to enter the date and time.  I suggest entering the date so that when you save a file, it will be stored with the current date on it.  When you enter the date, enter it using the same format as the date displayed.

If you are not interested in entering the time, just press the return key in response to the time request.

### 2.  PROTECTING YOUR DISKETTES

The first thing you should do is to make back-up copies of the testmaking diskettes.  Then use the copies when making tests or creating new test questions and store the original diskettes in a safe place.

In order to copy a diskette you need to have two newly formatted diskettes.  Formatting a diskette prepares it for referencing by the operating system and also erases any programs that might be stored on it.  Be sure you either use new diskettes or diskettes that no longer have programs you are interested in keeping.

## FORMATTING A DISKETTE

To format a diskette with one disk drive, put your operating system diskette in the drive and type:

format /s

and then press the return key (the key to the right of the keyboard that has a back arrow with a 90 degree bend in it). If you have two disk drives, put the operating system diskette in drive a: and put the diskette to be formatted into drive b:. Then type:

format b: /s

and press the return key.

The "/s" tells the computer to add system files to the diskette as well as format it. The three system files (COMMAND.COM and two small hidden files) are handy to have on your diskettes so you don't have to keep putting in a DOS diskette every time you change programs.

In both cases you will be prompted as to what to do next. More detailed instructions on formatting are found in the DOS operating system manual.

## COPYING A DISKETTE

Once you have two diskettes formatted, you are ready to copy the programs and test bank onto them.

If you have one drive, put the operating system into the drive and type:

diskcopy (return)

Then using either a newly formatted or an unformatted diskette, follow the prompts.

Once you have copied a diskette using "diskcopy", you should check to see that the copied diskette is the same as the original diskette. To do this, again put the operating system diskette in drive a: and type:

diskcomp (return)

Again follow the prompts as to when to switch diskettes.

If you want to put the systems files on either diskette, put your newly copied diskette into the drive and type:

sys (return)

If you have two disk drives, put the diskette to be copied in drive a: and the formatted diskette into drive b:. Then type:

copy a:*.* b: /v (return)

Prompts will tell you which diskette to put in which drive.

For more detailed instructions on copying files, see the DOS manual.

## 3.  SEEING WHAT'S ON A DISKETTE

Once your diskettes are copied, you should look at the file names on
each diskette.  To do this, put a diskette into drive a: and type:

                        dir (return)

On the program diskette, you should see the programs:

                        MAKETEST EXE
                        EDITBANK EXE

and possibly some chapter files.  Each chapter file is labeled by the
letters "CHPT", the chapter number, such as "4a", and then the extension
".MKT".  The chapter file name for chapter 4a looks like:

                        CHPT4A   MKT

When entering a file name plus the extension, you need to use a period
before the extension.  When the computer lists file names using the
"dir" command, it doesn't print the period.

If you have single sided drives, some of the files from the last
chapters in the book may be on your program diskette.  Most of the
chapter files will be on the question diskette.


## 4.  LOADING A PROGRAM

Both MAKETEST and EDITBANK have been compiled, so all you need to do in
order to run them is to put them into one of the drives and type the
program name (either MAKETEST or EDITBANK), followed by pressing the
return key.  The program will load into the computer and immediately
start executing.

If you have one disk drive, put the program diskette into the drive and
type the program name followed by pressing the return key.  For example,
if you want to use MAKETEST, type:

                        maketest (return)

Later, when the program asks for test bank questions, you will need to
take the program diskette out of the drive and replace it with the test
bank diskette.

This procedure will also work using the default drive if you have two
disk drives.

If you have two disk drives you can use both drives when  loading and
running a program.  To do this, first be sure drive a: is the default
drive by typing:

                        a: (return)

Then put the program diskette into the drive b: and type:

b:maketest (return)

Later, when the program asks you for it, put the question diskette in drive a:. All editing and test making will automatically be done using the diskette in the default drive, drive a:.

If for some reason a program quits working, the computer will return to the operating system and you will again get the A> prompt. To start the program again, follow the steps outlined above.

## 5. LISTING A CHAPTER FILE

In order to choose specific questions for a test, you need to have a listing of all the questions available. There are several ways you can get this.

The easiest way to see what questions are available is to look in the part F, "The Test Bank", of this booklet.

If you have added questions of your own to the test bank, they will not be in the booklet. In this case, you can get a compressed listing of questions in a chapter by using the "type" command in the operating system.

To list a file on the printer using DOS, put the question diskette in the default drive. Turn on the printer, making sure it is on line to the computer. Then press the "Ctrl" key (left side of keyboard) and "PrtSc" key (right side of keyboard) at the same time. Nothing will happen on the screen but the computer will set things up so that whatever appears on the screen will also be sent to the printer. Then to list a chapter, say chapter 4a, type the command:

type chpt4a.mkt (return)

(See section B3, "Seeing What's on a Diskette", for an explanation of chapter file names.) The computer will print the test bank file that corresponds to chapter 4a on the screen, and also on the printer.

The listing will look different from the booklet because it includes a code for each question in the chapter. A sample listing with codes is shown and explained in Appendix A.

Once you are familiar with MAKETEST, there is another way to print a chapter file. You can use the create option in MAKETEST to make a test that has every question in the chapter file on the test. This will give you an easy-to-read listing of questions and answers in a chapter file.

## 6. SEEING HOW MUCH SPACE IS ON A DISKETTE

Before using the program, EDITBANK, you should check to see how much space is on a diskette that you want to add to. If there is not enough space to add more questions to that diskette, you need to have a newly formatted one available.

The way to check a disk is to run the program called CHKDSK located on your operating system diskette. To run this program, put your operating diskette into drive a:, the disk to be checked into drive b: and type:

                    chkdsk b: (return)

The computer will check over the directory of the diskette in drive b: and display a staus report. If the diskette has more than 5000 or 10,000 bytes free, you can go ahead and use that diskette for storing new questions. If the reprort shows a lower number of bytes free, then you should refer to the section called "Formatting a Diskette" above and have a newly formatted diskette ready to store your questions.

# Running MakeTest

To start creating a test, boot up the computer and load MAKETEST according to the instructions above. Once you type the program name, MAKETEST should load and start executing.

## 1. STARTING MAKETEST

The first thing to appear is the title page with credits and copyright notice. When you are ready to begin, press any key (the spacebar is probably the easiest key to press).

Next MAKETEST will prompt you to put the question diskette into the default drive. The default drive is the drive that appears in the DOS prompt (e.g. A>). Unless you changed the default drive, it will still be drive a:. If necessary, take the program diskette out of drive a: and replace it with the question diskette. Be sure to close the drive door, then press any key to continue.

## 2. THE MAKETEST MENU

The main menu for MAKETEST looks like this:

    C.  Create test.

    V.  View test.

    P.  Print test.

    R.  Reorder test.

    X.  Exit from program.

In order to view, reorder or print a test, you must first create it. Since this version of MAKETEST will not store a created test, every time you start the program you will need to create a test. Once a test has been created, you can view, reorder or print it as many times as you want.

So press the "c" key. Later when you want to choose one of the other options, push the key that corresponds to that option. There is a double check on the exit key in case you accidentally push it.

Each of these options is explained in the following sections.

## 3. CREATING A TEST

Once you press the letter "c" in the main menu, a second menu will appear on the screen. This is the create-test menu and it looks like this:

    U.  You choose questions.

    C.  Computer chooses questions at random.

    X.  Exit to main menu.

Each of the first two options will create a test. The difference between these options is that in the "u" option, you choose every question on the test. In the "c" option, the computer will choose questions at random from the test bank.

The "X" option, is used in case you accidentally entered the Create Option from the main menu.


## CHOOSING SPECIFIC QUESTIONS

When choosing questions for the test, you will need to have a listing of questions available (See the section called "Listing a Chapter File" above for ways to get a listing). Once you know what questions are in a chapter file, and are ready to choose from those questions, press the "u" key.


### Choosing the Chapters

The computer will first ask which chapters you want on the test. Enter the number of each chapter you want, one chapter number at a time, each number followed by pressing the return key. As you enter chapter numbers, your choices will be shown at the bottom of the screen.

In general, questions from chapters 2 through 42 are available to use. The first 36 chapters correspond to chapters in "Conceptual Physics." The last six chapters (37 through 42) are chapters that you may have filled with questions of your own through using the EDITBANK program.

Sometimes the question diskette that has questions from a chapter you've asked for isn't in the default disk drive. When that happens, MAKETEST will try to access that file, but won't be able to find it and will print an error message that looks like this:

    CHPT2.MKT file not found.

At this point you have three choices. The first choice is to put the correct diskette into the default drive and press any key to continue making the test.

The second choice is to skip that chapter by pressing the "Alt" and "s" keys at the same time and continue making a test which will not include the skipped chapter.

The last choice is to return to the main menu by pressing the "Alt" and "x" keys at the same time. Doing this will erase any chapter choices you have made.

If you enter anything other than a number or if you enter a chapter number larger than 42, the computer will print a helping message.

When you finish entering chapter numbers press the return key without typing anything else on that line.

## Choosing the Questions

When the program asks for your question choices, type questions numbers that correspond to numbers in the test bank booklet (or in a listing, if you added questions of your own.). If you enter a question number that is larger than the total number of questions available, the computer will print a warning message. Again, your question choices will appear at the bottom of the screen.

When you finish entering specific question numbers, press the return key. The program will return to the main menu.

## LETTING THE COMPUTER CHOOSE QUESTIONS AT RANDOM

Once you press the letter "c" , the computer will get ready to choose questions from chapter files randomly within a given difficulty level.

## Choosing the Chapters

The first thing that happens is that the computer will ask for chapter numbers that you want on the test. Enter each chapter number one at a time, followed by pressing the return key. When you have finished entering chapter numbers, press the return key again.

In general, questions from chapters 2 through 42 are available to use. The first 36 chapters correspond to chapters in "Conceptual Physics." The last six chapters (37 through 42) are "empty" chapters that can be filled with questions of your own by using the EDITBANK program. Later in the program, if the computer can't find a chapter file, it will print a message on the screen and offer several options for continuing (see the section called "Choosing the Chapters" on the previous page for more information about this topic).

## Choosing the Questions

The computer will show you how many questions are available for the first five chapter choices and ask how many of these questions you want on the test.

Before responding to this part of MAKETEST, you need to know how questions are grouped.  Test bank questions are grouped in two different ways.  First, they are ranked A, B, or C, according to difficulty.  An A question is the easiest type of question, a B question the next easiest type, and a  C questions the hardest type of question.  Secondly, questions are grouped according to whether or not there is a mathematical calculation in them.  Altogether, there are six catagories - A, B, and C level conceptual questions, and A, B, and C level mathematical questions.

After you decide how many of each type of question you want in a given chapter, type in each amount as it is asked for. Your choices will be displayed at the bottom of the screen.

After the question amounts are entered, the computer gets questions from the test bank and stores them. At this point you need to be sure that the test bank question diskette is in the default drive and the drive door closed.

Once this part of the program is ended, it returns to the main menu.


## ESCAPE FROM CREATE TEST OPTION

If you accidentally enter the Create Test option, you can exit from that option by pressing the "x" key instead of either the "u" or "c" keys.  This will cause the program to return to the main menu without destroying any test you have already created.


## 4. VIEWING A TEST

Use this option to look at a test you have created.  It is useful for viewing answers and seeing which questions the computer has chosen.  If after viewing a test, you don't want that particular test, you can return to the Create Test option and create a different test.

The computer will display two questions at a time on the screen.  When you are ready to see the next two questions, push any key. After the entire test has been viewed, MAKETEST will print the answers to the test.

Test question answers are displayed according to the test questions they correspond to. The chapter number and test bank question number are also shown for easy referencing. A sample printout is shown here:

<pre>
        1      a     4a      3
</pre>

1    corresponds to question 1 on the test.
a    is the answer.
4a   is the chapter the question is stored in.
3    is the question number in that chapter file.

In this case, the first question on the test has an answer "a" and is question #3 in the chapter 4a test bank file.

The viewing option is not an editor - any typing you do here will have no effect on the test printout.

After all the answers have been viewed, the program returns to the main menu.

## 5. PRINTING A TEST

This is the most important part of MAKETEST - it prints the test you create.

The first thing to do is to type the test title when the program asks for it. Use only one line and press the return key when done.

After it gets the title, MAKETEST will ask for special instructions. These instructions will be printed at the top of the test just after the title. You may use up to 11 lines for the instructions.

When you type instructions, type one line at a time, pressing the return key at the end of each line. If you type more than 60 characters on a line, MAKETEST will erase the line and ask you to retype it.

After typing the title and special instructions, you need to be sure the printer is turned on and is on-line to the computer.

As the test is printed, MAKETEST will number questions in order. On a separate page at the end, it will print answers to the questions and where, in the test bank, the questions come from.

If you want to momentarily stop printing the test, you can do this by pressing the "Alt" and the "s" keys at the same time. To start printing again, press the "s" key or, to return to the main menu, press the "x" key.

MAKETEST will print a maximum of 80 characters per line, at a spacing of 6 lines to the inch. You can change the type of print that the printer uses by giving the printer some commands in BASIC before running MAKETEST. To do this, you need to know codes that your printer will understand. These codes are called "escape codes" and are generally found in the printer manual. Appendix B explains more on how to do this.

When MAKETEST has finished printing the test it will return to the main menu.


## 6. REORDERING A TEST


This section is used to reorder a test that has already been created through the Create Test option.  It will keep the same questions on the test, but reorder them based on a random number generator.  It reorders the entire test and will not reorder questions within a given chapter or difficulty level.

Sometimes, with a very few questions, reordering a test will give you the same test as before. This occurs because the reodering process is based on generating random numbers and with only a few numbers there is a high probability that the same sequence will appear twice in a row. If this happens, keep reordering the test until you get a new order.

Once the test is reordered, MAKETEST returns to the main menu.  Because MAKETEST is in a compiled form, this part of the program can run very fast.  Don't be surprised if you completely miss seeing the reordering message!

You can keep reordering or viewing a test as many times as you want before printing it.


## 7. ENDING THE SESSION


When you have finished using MAKETEST you can either press the "x" key in the main menu or just turn off the computer. Doing either one of these things will erase any test you have created, so be sure you have printed everything you want to keep.

# Running EditBank

EDITBANK is intended to allow you to add extra questions to the test bank and to edit those questions.

Before running EDITBANK, you should check each diskette to be sure there is enough space to store your questions. (See the section called "Seeing How Much Space is on a Diskette" above.)

## 1. STARTING EDITBANK

The first thing to do is to load EDITBANK into the computer as explained in the section called "Loading a Program" above.

Once EDITBANK is loaded and starts running, put the diskette with the test bank questions you will be editing into the default drive. Be sure to close the drive door when you are done.

## 2. THE EDITBANK MENU

The main menu for EDITBANK looks like this:

        C.   Create new file.

        A.   Add to existing file.

        E.   Edit existing file.

        U.   Update edited file.

        X.   Exit from program.

You can create chapter files of your own (chapters 37 through 42 are reserved for this), add to existing chapter files, or edit any of the files, including the ones that are in the test bank.

When editing or adding to an existing chapter file, you need to call the chapter by its number.  There are questions from chapters 2 through 36 (chapter 36 corresponds to Appendix V).  Some chapters, such as chapter 3 have two or three sections to them.  These sections are labeled with a lower case letter after the chapter number, such as 3a or 3b.  You need to look in the test bank booklet to see which chapters are sectioned. The program already has a list of the chapters and sections and will direct you to enter an existing chapter or subchapter.

If you want to create your own chapters, use chapters 37 through 42 to do this. None of these chapters are sectioned.

Once a question is created or edited, EDITBANK will make up a code line that has the chapter code, the question number, the number of lines in the question, the answer, the level of difficulty, and the question number within the difficulty level. Before returning to the main menu, EDITBANK will add this code to each question and store the question in order in the chapter file. Because of this it is important that you always return to the main menu before stopping.

ALWAYS RETURN TO THE MAIN MENU BEFORE ENDING A SESSION WITH EDITBANK.

Each of the EDITBANK options is explained in the following sections.

## 3. CREATING A NEW FILE

If you want to create a new chapter file, push the "c" key while in the main menu.

### Entering the Chapter Number

The first thing EDITBANK does is to ask you which chapter you want to create. There are six "empty" chapters, 37 through 42 that you can use. These chapters are not subdivided into sections. Enter a number between 37 and 42.

### Entering the Questions

Once you've told the computer which chapter you want to create, EDITBANK will print directions for entering questions. At the top of the screen, EDITBANK will show you which chapter you are working in, and which question you are entering.

When you enter a question, enter each line exactly as you want it to appear on the test. This means ending the line by pressing the enter key where you want to break it. You can use the back arrow, cursor right, cursor left, insert, and delete keys to edit a line before you press the enter key. All other cursor keys are disabled and may beep when pressed.

Be sure a line is correct before pressing the return key because once
you press the return key, the line is entered and you can change it only
by editing it.

When you enter multiple choice or true false answers, type in as many
spaces between parts of the answer as you want.  I suggest starting the
answers at space 5 and if there's a second answer, at space 29.  I also
suggest ending each line at or before the 60th character.  There are
helping dots located on the screen at each of these locations. For
example, a sample entry might look like:

Sound waves can interfere with one another so that no sound
results.
     a. true.                 b. false.

The "a" is lined up with the fifth space, the "b" with the 29th space
and the first line of the question ends in the 59th space.

Notice also, that answer choices are entered on the next line after the
question with no blank line between them and the question.  The
testmaking program, MAKETEST, will automatically add a blank line before
the first answer choice.

In order for MAKETEST to insert a blank line before answer choices, the
choices must be either letters starting with the letter "a" or numbers
starting with the number "1".

A question, plus answer choices, can be up to 11 lines long.  The
computer will not accept additional lines.

Once you finish entering a question, press the return key.  This lets
EDITBANK know you have finished.

### ENTERING THE ANSWER

Once the question is typed in, the program will ask you for an answer.
Enter the letter or number of the choice that is the correct answer.
For example in the above sample question the answer is "a". To enter the
answer letter, press the letter then press the return key.

### ENTERING THE DIFFICULTY LEVEL

After the answer is entered, EDITBANK will ask you for the difficulty
level.  Questions are grouped according to three levels of difficulty
(A, B, or C with A being the easiest level) and also according to
whether or not a question has a mathematical calculation (conceptual or
mathematical).  Altogether there are six levels of difficulty:

     A-conceptual, B-conceptual, C-conceptual,
   A-mathematical, B-mathematical, and C-mathematical.

You need to decide which level of difficulty your problem is and enter the number corresponding to that level. The small table shown above is printed on the bottom of the screen for reference when choosing the difficulty level. For example in the problem above, the difficulty level is A-conceptual, so you would enter the number 1.

## CLOSING THE FILE

Once you have finished entering questions type ENDFILE, in either upper or lower case letters. EDITBANK will save what you have typed and update the directory of how many questions are available on each level in the chapter. It is very important to allow the program to run through the ending and updating procedures. DO NOT BREAK THE PROGRAM BEFORE ALLOWING IT TO RETURN TO THE MAIN MENU.

## 4. ADDING TO A FILE

EDITBANK will first ask you which chapter you want to add to. If you want to add questions to a chapter, type the chapter number then the subchapter letter, a, b, or c, if necessary. Type the number exactly as it appears in the test bank booklet. For example to add questions to the chapter on "Projectiles" the chapter number would be "3a".

Once you have entered the chapter number, EDITBANK puts your entry at the end of the file. For example if there are 16 questions in chapter 3a, EDITBANK starts your entries with question 17.

The chapter number and question number appear at the top of the screen.

## ENTERING THE QUESTIONS

Enter questions to a chapter file exactly as you want them to appear on the test. See the section called "Entering the Questions" under the section called "Creating a New File" above for further instructions.

## ENTERING THE ANSWER

Enter the answer when asked for it.

## ENTERING THE DIFFICULTY LEVEL

Enter the difficulty level when asked for it. See the section called "Entering the Difficulty Level" above for further instructions.

## CLOSING THE FILE

Be sure to return to the main menu before exiting EDITBANK. See the section called "Returning to the Main Menu" above for more information on closing a file.

## 5. EDITING A FILE

Once EDITBANK returns to the main menu, you can edit the questions you just typed, or any of the test bank questions by pressing the "e" key.

EDITBANK will first ask for the chapter number that you want to edit. Type the chapter number including the subchapter letter if necessary. See the section called "Choosing the Chapters" under "running MAKETEST" above for more details on entering chapter numbers.

## CHOOSING TO EDIT A QUESTION

Once in the edit routine, EDITBANK will print each question in a chapter file on the screen. You have the options of skipping over, editing, deleting the question, or returning to the main menu.

If you want to skip over a question, press the return key.

If you want to edit the question, press the "Alt" and "e" keys at the same time.

If you want to delete a question, press the "Alt" and "d" keys at the same time.

If you want to return to the main menu, press the "Alt" and "x" keys at the same time.

If you want to edit a particular question, keep pressing the return key until the question you want appears. Then to edit the question, press both the "Alt" key and the "e" keys at the same time.

## EDITING A QUESTION

This part of the program is basically a line editor. Each question in a chapter file is shown along with the answer and difficulty level. When you indicate you want to edit a question, EDITBANK will step through the question line by line. Your choices are skipping a line, inserting a line before the current line, deleting a line or editing a line.

If you want to skip a line press the return key.

If you want to insert a line press both the "Alt" and "i" keys at the same time, then type the new line. Before pressing the return key, use the left and right cursor keys, back arrow, insert, and delete keys to edit the line. When the line is exactly the way you want it, press the return key.

In order to add a new line to the end of a question, you need to copy the last line of the question by inserting it before itself, insert the new last line, and then delete the old last line of the question.

If you want to delete a line, press both the "Alt" and "d" keys at the same time.

If you want to edit a line, press the "Alt" and "e" keys at the same time. The cursor will then appear on the screen at the beginning of the line. You must retype the entire line and then use any of the editing keys to make changes in the line. When you are finished, press the return key.


## CHANGING THE ANSWER

When you have finished editing a question, EDITBANK will ask if the answer is the same as it was. If the answer is different than it was before editing, press the "n" key, otherwise pressing any other key will indicate the answer is the same as it was.

If you are changing the answer, enter the new answer when asked for it.


## CHANGING THE DIFFICULTY LEVEL

Once you finish entering the answer, EDITBANK will ask if the difficulty level is the same as it was. Again, if it is different, press the "n" key, otherwise pressing any other key will indicate the difficulty level is the same as it was.

## CLOSING THE FILE

When you have completely finished with a question, EDITBANK will print the next question on the screen. If you are finished editing questions, you can return to the main menu by pressing both the "Alt" and the "x" keys at the same time. It is important that you press the "Alt" and "x" keys at the same time when done. Doing this saves what editing you have done, and updates the directory of questions available.

## 6. UPDATING A FILE

Since the editor part of this program is only a line editor, it is not the easiest way to edit a file. If you have an editing program that you use, you can use that program plus the "Update" option of EDITBANK to edit test bank files.

## USING A DIFFERENT EDITOR

If you use a different editor, you will need to know where the answer and the difficulty level are in the question code. The question code is explained in Appendix A.

If you have any changes to make to either the answer or difficulty level, you must make the changes in the code. You don't need to change anything in the code except possibly the answer or difficulty level.

The "Update" routine will automatically recount line numbers and renumber questions.

To edit questions using a text editor, edit the text part of a file and save the edited file on the question diskette. Then use the "Update" option in EDITBANK. This routine revises and correctly enters codes at the beginning of each question and will also update the directory of questions available which is stored at the beginning of every chapter file.

Since every text editor works a little bit differently, I suggest you try editing a meaningless file to see if your editor plus the Update routine will work together.

If you use a separate editor, it is necessary to use the "Update" option before running MAKETEST.

USING THE UPDATE OPTION

To use the Update option, press "u" when in the main menu.  When the
program asks for which file to update, enter the number of the file you
have edited.  For example, if you edited chapter 3a, then type:

                                    3a

in response to that question.

Be sure to return to the main menu before exiting from EDITBANK or
turning off the computer.


## 7. ENDING THE SESSION


When you are completely finished using EDITBANK, press the "x" key while
in the main menu.  The program will return to the operating system.

# Appendices

1. APPENDIX A
(Question Code)


If you list a chapter file, you will see that it looks slightly
different from a test.  At the beginning of every question there is a
question code that contains relevant information about the problem.
This section shows a short listing for chapter 3a and an explanation of
the question codes.


Chapter 3a
(Shortened Listing)


8
3
2
1
0
0
2
>>>>020106c  101
A bullet fired horizontally hits the ground in 0.5 seconds.
If it had been fired with a much higher speed in the same
direction, it would have hit the ground (neglect the
Earth's curvature) in
    a. less than 0.5 s.    c. 0.5 s.
    b. more than 0.5 s.
>>>>020205c  102
A projectile is fired straight up into a vacuum at a speed
of 100 m/s.  The projectile returns to its original
starting position at a speed of
    a. less than 100 m/s.    c. 100 m/s.
    b. more than 100 m/s.
>>>>020307b  103
A projectile is fired horizontally.  The projectile
maintains its horizontal component of speed because it
    a. is not acted on by any forces.
    b. is not acted on by any horizontal forces.
    c. has no vertical component of speed to begin with.
    d. the net force acting on it is zero.
    e. none of these.
>>>>020406b  201
An object is dropped and freely falls to the ground with an
acceleration of 1 g.  If it is thrown upward at an angle
instead, its acceleration would be
    a. 0 g.              d. larger than 1 g.
    b. 1 g downward.    e. none of these.
    c. 1 g upwards.

>>>>020406a 202
A projectile is hurled into the air at an angle of 50
degrees and lands on top of a target. It will also land on
top of the target if it is thrown at an angle of
    a. 40 degrees.              d. 60 degrees.
    b. 45 degrees.              e. none of these.
    c. 55 degrees.
>>>>020606a 301
An object is thrown vertically into the air. Because of air
resistance, the time for its decent will be
    a. longer than the ascent time.
    b. shorter than the ascent time.
    c. equal to the ascent time.
    d. not enough information given to say.
>>>>020707c 601
A ball player wishes to determine her or his pitching speed
by throwing a ball horizontally from an elevation of 5 m
above the ground. The player sees the ball land 20 m down
range. What is the player's pitching speed?
    a. 5 m/s.                   d. 25 m/s.
    b. 10 m/s.                  e. none of these.
    c. 20 m/s.
>>>>020808a 602
An airplane flies at 40 m/s at an altitude of 50 meters.
The pilot drops a package which falls and strikes the
ground. Where, horizontally, does the package land?
    a. right beneath the plane.
    b. 400 m behind the plane.
    c. 500 m behind the plane.
    d. more than 500 m behind the plane.
    e. none of these.
ENDFILE


THE FIRST NUMBERS


The numbers at the beginning are the directory for this shortened
version of the chapter 3a file.

The first number, 8, is the total number of questions in the chapter
file.

The second number, 3, is the number of questions at difficulty level 1
in this file.

The third number is the number of questions at difficulty level 2 and so
on.

For example, this chapter file has 2 questions at the hardest difficulty
level, 6.

## THE CODE

The code for the third problem looks like this:

>>>>020307b   103

>>>>   is a signal that a new question is beginning;
02    is the chapter number code recognized by the computer (Chapter 3b
      has code 03);
03    means this is the third question in the chapter file;
07    means there are 7 lines of question, including the answer choices;
b     is the answer;
1     is the code for the difficulty level A, conceptual; (the codes for
      each difficulty level are listed below)

> 1 is A level, conceptual,
> 2 is B level, conceptual,
> 3 is C level, conceptual,
> 4 is A level, mathematical,
> 5 is B level, mathematical,
> 6 is C level, mathematical;

03    means this is the third question of difficulty level 1 in this
      chapter.


## THE REST OF THE FILE

Each question follows its question code.

At the end of the file, the word, ENDFILE (in capital letters), is
printed. This is a signal to the program that the chapter file has
ended.

If you edit a chapter file using another editor, it is very important to
be sure the ENDFILE statement is the last line in the file.  It needs to
be typed in capital letters.


## 1. APPENDIX B
## (Printer Controls)

It is possible to print a test using different modes of printing. For
example you can command the printer to print each line darker, or to
compress the test onto the left half of a page.  You need to look at
your printer instruction manual to see exactly which modes are available
to you and also to see what the codes are for each mode.

Once you know which mode you want to use, and what its code is, then you
can command the printer to print in that mode by using the BASICA
interpreter.

For example, suppose you want to use double strike mode on the printer.
(This mode scrolls the paper down a tiny bit and types over a line.)
To do this, put your system diskette into the default drive and type:

basica (return)

When the interpreter is loaded and running, turn on the printer,
making sure it is on-line to your computer, and type

lprint chr$(27) chr$(71)

pressing the return key at the end.

On an IBM printer, chr$(27) stands for the character whose ASCII code is
27.  This is the "escape" character.  Most computer codes will use the
"escape" character as part of the code.  The second word, chr$(71),
stands for the character whose ASCII code is 71.  This character is a
"G".  The entire code, "escape G" tells the printer to use double strike
mode when printing.

The printer will continue to use double strike mode until either you
tell it to stop using that mode or until the printer is turned off.  The
code to tell the printer to stop using double strike is:

lprint chr$(27) chr$(72)

which stands for "escape H".

You can use any combination of printer codes, but you need to remember
which ones you are using, and which ones you might want to turn off.

Some printer codes for IBM and Epson printers are listed below.  These
may or may not work on your printer.  Try them!

| "shift in" | chr$(15) | turns on compressed mode (will not combine with emphasized mode) |
|------------|----------|------------------------------------------------------------------|
| "DC2" | chr$(18) | turns off compressed mode |
| "escape E" | chr$(27) chr$(69) | turns on emphsized mode |
| "escape F" | chr$(27) chr$(70) | turns off emphasized mode |
| "escape G" | chr$(27) chr$(71) | turns on double strike mode |
| "escape H" | chr$(27) chr$(72) | turns off double strike mode |

These next two codes are for Epson printers only:

| "escape 4" | chr$(27) chr$(52) | italics font on |
|------------|-------------------|-----------------|
| "escape 5" | chr$(27) chr$(53) | italics font off |

If you turn the printer off, it will forget any codes you have entered
and will return to the default mode which is standard type, single
strike print.

Once you finish setting the mode, return to the operating system, by
typing

system (return),

and load MAKETEST as described in part C above.

# TEST BANK

Paul G. Hewitt
City College of San Francisco

SUMMARY OF QUESTIONS

| CHAPTER | DIFFICULTY LEVEL | | | | | | |
|---|---|---|---|---|---|---|---|
| | A | B | C | A(M) | B(M) | C(M) | Total |
| Part 1: Mechanics ...................................................... | | | | | | | |
| 2: Linear Motion | 10 | 7 | 3 | 4 | 9 | 3 | 36 |
| 3a: Projectiles | 5 | 6 | 1 | 0 | 0 | 0 | 14 |
| 3b: Satellite Motion | 3 | 14 | 0 | 0 | 0 | 0 | 17 |
| 4a: Non-Free Fall | 5 | 6 | 3 | 0 | 0 | 3 | 17 |
| 4b: Newton's 1st & 2nd Laws | 20 | 17 | 3 | 2 | 2 | 9 | 54 |
| 4c: Newton's Third Law | 4 | 9 | 1 | 0 | 0 | 2 | 16 |
| 5: Momentum | 13 | 9 | 6 | 3 | 1 | 0 | 32 |
| 6a: Work and Energy | 9 | 8 | 2 | 8 | 8 | 6 | 41 |
| 6b: Momentum and Energy | 9 | 2 | 4 | 0 | 0 | 3 | 18 |
| 7: Rotational Motion | 12 | 7 | 9 | 0 | 0 | 2 | 30 |
| 8a: Gravitation | 6 | 12 | 5 | 0 | 0 | 0 | 23 |
| 8b: Tides | 1 | 2 | 5 | 0 | 0 | 0 | 8 |
| Part 2: Properties of Matter ..................................................... | | | | | | | |
| 9: Matter | 22 | 2 | 0 | 0 | 0 | 0 | 24 |
| 10a: Solids | 11 | 4 | 0 | 1 | 2 | 0 | 18 |
| 10b: Scaling | 14 | 5 | 0 | 0 | 0 | 0 | 19 |
| 11a: Liquids | 12 | 11 | 16 | 0 | 0 | 0 | 39 |
| 11b: Pascal's Principle and Surface Tension | 8 | 1 | 0 | 0 | 0 | 0 | 9 |
| 12a: Gases | 13 | 18 | 2 | 0 | 0 | 0 | 33 |
| 12b: Plasmas | 7 | 0 | 0 | 0 | 0 | 0 | 7 |
| Part 3: Heat ..................................................... | | | | | | | |
| 13: Temperature, Heat and Expansion | 7 | 11 | ? | 0 | 0 | ? | ?? |
| 14: Heat Transfer | 4 | 8 | 3 | 0 | 0 | 0 | 15 |
| 15: Change of State | 10 | 16 | 0 | 0 | 0 | 0 | 26 |
| 16: Thermodynamics | 11 | 17 | 0 | 1 | 1 | 2 | 32 |
| Part 4: Sound ..................................................... | | | | | | | |
| 17: Vibrations and Waves | 6 | 11 | 8 | 3 | 4 | 0 | 32 |
| 18: Sound | 11 | 8 | 2 | 0 | 1 | 3 | 25 |
| 19: Music | 8 | 11 | 5 | 3 | 2 | 5 | 34 |

| Chapter | A | B | C | A(M) | B(M) | C(M) | Total |
|---|---|---|---|---|---|---|---|
| Part 5: Electricity and Magnetism ..................................... | | | | | | | |
| 20: Electrostatics | 6 | 14 | 6 | 0 | 4 | 0 | 30 |
| 21: Electric Current | 8 | 1 | 5 | 3 | 2 | 5 | 34 |
| 22: Magnetism | 10 | 4 | 2 | 0 | 0 | 0 | 16 |
| 23: Electromagnetic Interactions | 4 | 4 | 3 | 0 | 3 | 0 | 14 |
| 24: Electromagnetic Radiation | 13 | 7 | 0 | 0 | 3 | 0 | 23 |
| Part 6: Light ..................................................... | | | | | | | |
| 25: Reflection & Refraction | 7 | 19 | 8 | 0 | 2 | 0 | 36 |
| 26: Color | 12 | 6 | 8 | 0 | 0 | 0 | 26 |
| 27: Interference and Polarization | 5 | 8 | 2 | 0 | 0 | 1 | 16 |
| 28: Light Emission | 8 | 11 | 2 | 0 | 0 | 0 | 21 |
| 29: Light Quanta | 8 | 5 | 1 | 0 | 0 | 0 | 14 |
| Part 7: Atomic and Nuclear Physics ................................... | | | | | | | |
| 30: Atoms and Quanta | 17 | 4 | 6 | 0 | 0 | 0 | 27 |
| 31: The Atomic Nucleus and Radioactivity | 18 | 11 | 3 | 0 | 4 | 0 | 36 |
| 32: Nuclear Fission & Fusion | 15 | 20 | 4 | 0 | 0 | 0 | 39 |
| Part 8: Relativity and Astrophysics .................................. | | | | | | | |
| 33: The Special Theory of Relativity | 11 | 13 | 4 | 0 | 0 | 0 | 28 |
| 34: The General Theory of Relataivity | 10 | 11 | 5 | 0 | 0 | 0 | 26 |
| 35: Astrophysics | 15 | 13 | 2 | 0 | 0 | 0 | 30 |
| Appendix ............................................................. | | | | | | | |
| Appendix 5: Exponential Growth | 3 | 1 | 0 | 0 | 3 | 1 | 8 |

# 2

# Motion

Chapter 2: Linear Motion

1. As an object freely falls, its  (a)A

   a. velocity increases.
   b. acceleration increases.
   c. both of the above.
   d. none of the above.

2. If a freely-falling object were somehow equipped with a  (b)A
   speedometer, its speed reading would increase each second
   by about

   a. 5 m/s.           d. a variable amount.
   b. 10 m/s.          e. depends on its
   c. 15 m/s.             initial speed.

3. If a freely-falling object were somehow equipped with a  (b)A
   speedometer on a planet where the acceleration due to
   gravity is 20 meters per second squared, then its speed
   reading would increase each second by

   a. 10 m/s.          d. 40 m/s.
   b. 20 m/s.          e. depends on its
   c. 30 m/s.             initial speed.

4. If a freely-falling object were somehow equipped with an  (c)A
   odometer to measure the distance it travels, then the
   amount of distance it travels each succeeding second would
   be

   a. constant.        c. greater than the
   b. less and less.      second before.

5. If an object falls with constant acceleration, the velocity  (b)A
   of the object must

   a. be constant also.
   b. continually change by the same amount each second.
   c. continually change by varying amounts depending on
      its speed at any point.
   d. continually decrease.
   e. none of these.

6. An object travels 8 meters in the first second of travel,  (a)A
   8 meters again during the second second of travel, and
   8 meters again during the third second.  Its acceleration
   is

   a. 0 meter per second squared.
   b. 8 meters per second squared.
   c. 16 meters per second squared.
   d. 32 meters per second squared.
   e. none of these.

7. An object is in free fall.  At one instant, it is traveling    (c)A
   at a speed of 50 meters per second.  Exactly one second
   later, its speed is about

    a. 32 m/s.        d. 88 m/s.
    b. 50 m/s.        e. 100 m/s.
    c. 60 m/s.

8. Disregarding air resistance, objects fall at constant    (c)A

    a. velocity.      d. distances each successive
    b. speed.           second.
    c. acceleration.  e. all of these.

9. A heavy object and a light object are dropped at the same    (b)A
   time from rest in a vacuum.  The heavier object reaches
   the ground

    a. sooner than the lighter object.
    b. at the same time as the lighter object.
    c. later than the lighter object.

10. A ball is thrown upwards and caught when it comes back    (c)A
    down.  Neglecting air resistance, the speed with which it
    is caught is

    a. more than the speed it had when thrown upwards.
    b. less than the speed it had when thrown upwards.
    c. the same as the speed it had when thrown upwards.

11. Starting from rest, a freely-falling object will fall    (d)AM
    in 10 seconds, a distance of about

    a. 10 m.        d. 500 m.
    b. 50 m.        e. more than 500 m.
    c. 100 m.

12. Ten seconds after starting from rest, a freely-falling    (c)AM
    object will have a speed of about

    a. 10 m/s.      d. 500 m/s
    b. 50 m/s.      e. more than 500 m/s.
    c. 100 m/s.

13. A car accelerates at 2 meters per second squared.  What is    (d)AM
    its speed 3 seconds after the car starts moving?

    a. 2 m/s.        d. 6 m/s.
    b. 3 m/s.        e. none of these.
    c. 4 m/s.

14. Ten seconds after starting from rest, a car is moving at    (c)AM
    40 m/s.  What is the car's acceleration?

    a. 0.25 meters per second squared.
    b. 2.5 meters per second squared.
    c. 4.0 meters per second squared.
    d. 10 meters per second squared.
    e. 40 meters per second squared.

15. When a rock thrown straight upwards gets to the exact top                          (b) B
    of its path, its

    a. velocity is zero and its acceleration is zero.
    b. velocity is zero and its acceleration is about
       10 meters per second squared.
    c. velocity is about 10 m/s and its acceleration
       is zero.
    d. velocity is about 10 m/s and its acceleration
       is about 10 meters per second squared.
    e. none of these.

16. A bullet is dropped from the top of the Empire State                                (c) B
    Building while another bullet is fired downward from the
    same place.  Neglecting air resistance, acceleration is
    greatest for the

    a. fired bullet.
    b. dropped bullet.
    c. ...is 9.8 meters per second squared for each.
    d. depends on how far they are above the ground.

17. A bullet is fired straight down from the top of a high                              (c) B
    cliff.  Neglecting air resistance, the acceleration of
    the bullet is

    a. less than 9.8 meters per second squared.
    b. more than 9.8 meters per second squared.
    c. 9.8 meters per seconds squared.
    d. more information needed to determine.

18. A package falls off a truck that is moving at 30 m/s.                               (c) B
    Neglecting air resistance, the horizontal speed of the
    package just before it hits the ground is

    a. zero.
    b. less than 30 m/s but larger than zero.
    c. 30 m/s.
    d. more than 30 m/s.
    e. more information needed to determine.

19. The muzzle velocity of a certain rifle is 100 m/s.  At                              (a) B
    the end of one second, a bullet fired straight up into
    the air will travel a distance of

    a. (100 - 4.9) m.    d. 4.9 m.
    b. (100 + 4.9) m.    e. none of these.
    c. 100 m.

20. If you drop an object, it will accelerate downward at a                             (c) B
    rate of 9.8 meters per second squared.  If you instead throw
    it downwards, its acceleration (in the absence of air
    resistance) will be

    a. greater than 9.8 meters per second squared.
    b. less than 9.8 meters per second squared.
    c. 9.8 meters per second squared.

21. A feather and a coin will have equal accelerations when                             (d) B
    falling in a vacuum because

    a. their velocities are the same.
    b. gravity does not act in a vacuum.
    c. the gravitational force on each is the same.
    d. the ratio of each object's weight to its mass is the same.
    e. none of these.

22. Starting from rest, a freely falling object will fall in     (c)BM
    0.5 seconds, a distance of about

      a. 0.5 m.            d. 5.0 m.
      b. 1.0 m.            e. none of these.
      c. 1.25 m.

23. One half second after starting from rest, a freely-falling     (c)BM
    object will have a speed of about

      a. 20 m/s.           d. 2.5 m/s.
      b. 10 m/s.           e. none of these.
      c. 5 m/s.

24. An object falls freely from rest on a planet where the     (e)BM
    acceleration due to gravity is 20 meters per second
    squared. After 5 seconds, the object will have a speed of

      a. 5 m/s.            d. 50 m/s.
      b. 10 m/s.           e. 100 m/s.
      c. 20 m/s.

25. An object falls freely from rest on a planet where the     (c)BM
    acceleration due to gravity is 20 meters per second
    squared. After 5 seconds it falls a distance of

      a. 100 m.           d. 500 m.
      b. 150 m.           e. none of these.
      c. 250 m.

26. An apple falls from a tree and hits the ground 5 meters     (b)BM
    below. It hits the ground with a speed of about

      a. 5 m/s.            d. 20 m/s.
      b. 10 m/s.           e. not enough information
      c. 15 m/s.                given to estimate.

27. It takes 6 seconds for a stone to fall to the bottom of a     (c)BM
    mine shaft. How deep is the shaft?

      a. about 60 m.       c. about 180 m.
      b. about 120 m.      d. more than 200 m.

28. A car accelerates at 2 meters per second squared. Assuming     (d)BM
    the car starts from rest, how far will it travel in
    10 seconds?

      a. 2 meters.         d. 100 meters.
      b. 10 meters.       e. 200 meters.
      c. 40 meters.

29. A car accelerates at 2 meters per second squared. Assuming     (b)BM
    the car starts from rest, how much time does it need to
    accelerate to a speed of 30 m/s?

      a. 2 seconds.       d. 60 seconds.
      b. 15 seconds.      e. none of these.
      c. 30 seconds.

30. A car accelerates from rest for 5 seconds until it reaches     (c)BM
    a speed of 22 m/s. What is the car's acceleration?

      a. 1.1 meters per seond squared.
      b. 2.2 meters per seond squared.
      c. 4.4 meters per seond squared.
      d. 5.0 meters per seond squared.
      e. 22 meters per seond squared.

31. A bullet is dropped into a river from a very high bridge.      (c)C
    At the same time, another bullet is fired from a gun,
    straight down towards the water.  Neglecting air resistance,
    the acceleration just before striking the water

        a. is greater for the dropped bullet.
        b. is greater for the fired bullet.
        c. is the same for each bullet.
        d. depends on how high they started.
        e. none of these.

32. Someone standing at the edge of a cliff throws one ball        (c)C
    straight up and another ball straight down at the same
    initial speed.  Neglecting air resistance, the ball to
    hit the ground below the cliff with the greatest speed
    will be the one initially thrown

        a. upward.              c. ...they will both hit
        b. downward.               with the same speed.

33. In each second of fall, the distance a freely falling         (d)C
    object will fall is about

        a. 5 m.
        b. 10 m.
        c. equal amounts, but not 5 m or 10 m.
        d. unequal amounts depending on the time of fall squared.
        e. none of these.

34. A ball is thrown upwards. What upward speed does the ball      (a)CM
    need to have to remain in the air for a total time of
    10 seconds?

        a. 50 m/s.              d. 100 m/s.
        b. 60 m/s.              e. 110 m/s.
        c. 80 m/s.

35. A ball is thrown 125 meters upward and then falls the          (b)CM
    same distance back to Earth.  Its total time in the air
    is about

        a. 5 seconds.           c. 15 seconds.
        b. 10 seconds.          d. more than 20 seconds.

36. A pot falls from a ledge and hits the ground 45 m below.       (a)CM
    The speed with which it hits the ground is about

        a. 30 m/s.              c. 120 m/s.
        b. 60 m/s.              d. more than 120 m/s.

# 3
# Projectile and Satellite Motion

Chapter 3a: Projectiles

1. A bullet fired horizontally hits the ground in 0.5 seconds.          (c)A
   If it had been fired with a much higher speed in the same
   direction, it would have hit the ground (neglecting the
   Earth's curvature and air resistance) in

   a. less than 0.5 s.      c. 0.5 s.
   b. more than 0.5 s.

2. A projectile is fired straight up into a vacuum at a speed          (c)A
   of 100 m/s.  The projectile returns to its original
   starting position at a speed of

   a. less than 100 m/s.    c. 100 m/s.
   b. more than 100 m/s.

3. A projectile is fired horizontally in a vacuum.  The               (b)A
   projectile maintains its horizontal component of speed
   because it

   a. is not acted on by any forces.
   b. is not acted on by any horizontal forces.
   c. has no vertical component of speed to begin with.
   d. the net force acting on it is zero.
   e. none of these.

4. A hunter aims a rifle at an angle of 10 degrees above the          (a)A
   horizontal.  The hunter fires a bullet while simultaneously
   dropping another bullet from the level of the rifle.  Which
   bullet will hit the ground first?

   a. the dropped one.      c. both hit at the
   b. the fired one.           same time.

5. A hunter aims a rifle at an angle of 10 degrees below the          (b)A
   horizontal.  The hunter fires a bullet while simultaneously
   dropping another bullet from the level of the rifle.  Which
   bullet will hit the gound first?

   a. the dropped one.      c. both hit at the
   b. the fired one.           same time.

6. An object is dropped and freely falls to the ground with an        (b)B
   acceleration of 1 g.  If it is thrown upward at an angle
   instead, its acceleration would be

   a. 0 g.                  d. larger than 1 g.
   b. 1 g downward.         e. none of these.
   c. 1 g upwards.

7. A projectile is hurled into the air at an angle of 50                    (a)B
   degrees and lands on a target that is at the same level the
   projectile started.  It will also land on the target if it
   is thrown at an angle of

      a. 40 degrees.      d. 60 degrees.
      b. 45 degrees.      e. none of these.
      c. 55 degrees.

8. A rifle with a muzzle velocity of 100 m/s is fired                        (c)B
   horizontally from a tower.  Neglecting air resistance, where
   will the bullet be 1 second later?

      a. 50 m down range.      d. 490 m down range.
      b. 98 m. down range.      e. none of these.
      c. 100 m down range.

9. Two projectiles are fired at equal speeds but different                   (a)B
   angles.  One is fired at an angle of 30 degrees and the
   other at 60 degrees.  The projectile to hit the ground
   first will be the one fired at (neglect air resistance)

      a. 30 degrees.      c. both hit at the
      b. 60 degrees.         same time.

10. After a rock that is thrown straight up reaches the top of               (c)B
    its path and is starting to fall back down, its
    acceleration is (neglect air resistance)

      a. greater than when it was a the top of its path.
      b. less than when it was at the top of its path.
      c. the same as it was at the top of its path.

11. In the absence of air resistance, a heavy boulder and a                  (d)B
    small rock dropped from the same height will hit the ground
    with identical speeds because

      a. the rock has less kinetic energy than the boulder.
      b. the rock has more kinetic energy than the boulder.
      c. the rock has the same kinetic energy as the boulder.
      d. none of these.

12. An object is thrown vertically into the air.  Because of air             (a)C
    resistance, the time for its descent will be

      a. longer than the ascent time.
      b. shorter than the ascent time.
      c. equal to the ascent time.
      d. not enough information given to say.

13. A ball player wishes to determine her or his pitching speed              (c)CM
    by throwing a ball horizontally from an elevation of 5 m
    above the ground.  The player sees the ball land 20 m down
    range.  What is the player's pitching speed?

      a. 5 m/s.      d. 25 m/s.
      b. 10 m/s.      e. none of these.
      c. 20 m/s.

14. An airplane flies at 40 m/s at an altitude of 50 meters.                 (a)CM
    The pilot drops a package which falls and strikes the
    ground.  Where, horizontally, does the package land?

      a. right beneath the plane.
      b. 400 m behind the plane.
      c. 500 m behind the plane.
      d. more than 500 m behind the plane.

Chapter 3b: Satellite Motion

1. The circular orbit of a satellite orbiting the Earth is          (d)A
   characterized by a constant

    a. speed.             d. all of the above.
    b. acceleration.    e. none of the above.
    c. radius.

2. The fastest moving planet in a solar system is                   (c)A

    a. the smallest planet.
    b. the most massive planet.
    c. the planet nearest the Sun.
    d. the planet farthest from the Sun.
    e. any planet- they all move at the same speed.

3. It takes Pluto a longer time to travel around the Sun           (c)A
   than the Earth does because Pluto

    a. has farther to go.    c. both of the above.
    b. goes slower.        d. none of the above.

4. An object is placed exactly halfway between the Earth and       (a)B
   Moon.  The object will fall toward the

    a. Earth.             c. neither of these.
    b. Moon.

5. A woman on the surface of the Earth has a mass of 50            (b)B
   kilograms and a weight of 490 newtons.  If the woman were
   floating freely inside a space habitat far away from Earth,
   she would have

    a. less weight and more mass.
    b. less weight and the same mass.
    c. less weight and less mass.
    d. more weight and less mass.
    e. none of these.

6. A "weightless" astronaut in an orbiting shuttle is             (d)B

    a. shielded from the Earth's gravitational field.
    b. beyond the pull of gravity.
    c. pulled only by gravitation to the shuttle which
       cancels the Earth's gravitational pull.
    d. like the shuttle, pulled by Earth's gravitation.
    e. none of these.

7. A projectile is fired vertically from the surface of the       (c)B
   Earth at 10 km/s.  The projectile will

    a. go into circular about the Earth.
    b. go into an elliptical orbit about the Earth.
    c. rise and fall back to the Earth's surface.
    d. none of these.

8. A vertically oriented rocket that maintains a continuous       (a)B
   upward velocity of 8 km/s will

       a. escape the Earth's g-field.
       b. not to be able to escape the Earth's g-field.


9. An object dropped from rest infinitely far from Earth falls    (c)B
   to the Earth because of the Earth's gravitational field and
   strikes the surface with a speed of about

       a. 9.8 m/s.            c. 11 km/s.
       b. 8 km/s.             d. none of these.


10. An Earth satellite is in an elliptical orbit.  The satellite  (a)B
    travels fastest when it is

       a. nearest the Earth.
       b. farthest from the Earth.
       c. ... it travels at constant speed everywhere in orbit.


11. Minimal orbit speed about the Earth is about 8 km/s.          (a)B
    Minimal orbital speed about the Moon would be

       a. less than 8 km/s.   c. about 8 km/s.
       b. more than 8 km/s.


12. The path of a satellite about a small planet is               (b)B

       a. hyperbolic.         c. parabolic.
       b. elliptical.         d. not enough information given.


13. A satellite in an elliptical orbit travels at                (e)B

       a. constant velocity.   d. all of the above.
       b. constant speed.      e. none of the above.
       c. constant acceleration.


14. The Earlybird communication satellite hovers over the same   (e)B
    point on Earth's equator indefinitely. This is because

       a. forces other than Earth's gravity act on it.
       b. it pulls as hard on the Earth as the Earth pulls on it.
       c. it is beyond the main pull of gravity.
       d. it is kept aloft by ground control.
       e. its orbital period is 24 hours.


15. A satellite near the Earth makes a full circle in about an   (a)B
    hour and a half. How long would a satellite located as far
    away as the Moon take to orbit the Earth?

       a. about 28 days.
       b. about an hour and a half.
       c. it depends on the mass of the satellite.
       d. a satellite could not be put into such an orbit.
       e. none of these.


16. Communications and weather satellites always appear at the   (c)B
    same place in the sky.  This is because these satellites
    are

       a. beyond the pull of the Earth's gravitational field.
       b. moving at a speed just short of escape velocity.
       c. orbiting the Earth with a 24 hour period.
       d. stationary in space.
       e. none of these.

17. Compared to the period of satellites in orbit close to the Earth,      (a)B
    the period of satellites in orbit far from the Earth is

    a. longer.              c. the same.
    b. shorter.             d. not enough information.

Chapter 4a: Non-Free Fall

1. Two objects of the same size, but unequal weights are                    (b)A
   dropped from a tall tower.  Taking air resistance into
   consideration, the object to hit the ground first will be
   the

       a. lighter object.    c. both hit at the same time.
       b. heavier object.    d. not enough information.

2. A light woman and a heavy man jump from an airplane at the               (a)A
   same time and open their parachutes at the same time.
   Which person will get to a state of zero acceleration first?

       a. the light woman.    c. both should at the same time.
       b. the heavy man.    d. not enough information.

3. A large and a small person wish to parachute at equal                    (a)A
   terminal velocities.  The larger person will have to

       a. get a larger parachute.
       b. jump lightly.
       c. pull upward on the supporting strands to decrease
          the downward net force.
       d. jump first from the plane.
       e. none of these.

4. A skydiver, who weighs 500 N, reaches terminal velocity                   (d)A
   of 90 kilometers per hour.  The air resistance the diver
   encounters in fall is

       a. 90 N.    d. 500 N.
       b. 250 N.    e. none of these.
       c. 410 N.

5. A sack of potatoes weighting 200 N falls from an airplane.                (a)A
   As the velocity of fall increases, air resistance also
   increases.  When air resistance equals 200 N, what is the
   sack's acceleration?

       a. 0 meters per second squared.
       b. 4.9 meters per second squared.
       c. 9.8 meters per second squared.
       d. infinite.
       e. none of these.

6. An elephant and a feather fall through the air.  The force                (a)B
   of air resistance is greater on the

       a. elephant.    c. is the same on
       b. feather.       each.

7. When an object falls through the air, its velocity          (b)B
   increases and its acceleration

        a. increases.           c. remains the same whether
        b. decreases.              in air or in vacuum.

8. A woman with a parachute on jumps from a high-flying plane.  (b)B
   As the woman's velocity of fall increases, her
   acceleration

        a. increases.           c. remains unchanged regardless
        b. decreases.              of air resistance.

9. A man falling through the air with a parachute on weighs     (d)B
   500 N.  When he opens his chute, he experiences an initial
   air resistance force of 800 N.  The net force on the man is

        a. 300 N downward.      d. 300 N upward.
        b. 500 N downward.      e. 500 N upward.
        c. 800 N downward.

10. An empty roller coaster car at an amusement park takes      (c)B
    3 minutes to make its ride from start to finish.
    Neglecting friction, a fully-loaded car would take

        a. less than 3 minutes.
        b. more than 3 minutes.
        c. 3 minutes.

11. A rock is thrown vertically into the air.  At the top of    (b)B
    its path, its acceleration is

        a. zero.
        b. 9.8 meters per second squared.
        c. between 0 and 9.8 meters per second squared.
        d. greater than 9.8 meters per second squared.
        e. none of these.

12. A ball is thrown vertically into the air at 20 m/s.         (a)C
    Because of air resistance, its speed when returning to its
    starting position will be

        a. less than 20 m/s.  c. more than 20 m/s.
        b. equal to 20 m/s.

13. A ball is thrown vertically into the air.  Because of air   (c)C
    resistance, the time taken in returning from the top to its
    starting level is

        a. less than the time taken going up.
        b. the same as the time taken going up.
        c. more than the time taken going up.

14. With no air resistance, a ball thrown straight up will land (b)C
    in 10 seconds.  In the presence of air resistance, a ball
    thrown at the same initial speed will land in

        a. less than 10 seconds.
        b. more than 10 seconds.
        c. 10 seconds.

15. A ball thrown straight upward takes 10 seconds to go up and (a)CM
    return to the ground.  Because of air resistance, the time
    taken for the ball just to go up is

        a. less than 5 s.      c. 5 s.
        b. more than 5 s.

- 40 -

16. A falling skydiver of mass 100 kg experiences 500 N air resistance.  What is his or her acceleration?                                    (d)CM

    a. 0.2 g.             d. 0.5 g.
    b. 0.3 g.             e. more than 0.5 g.
    c. 0.4 g.

17. An astronaut on another planet drops a 1 kg rock from rest. The astronaut notices that the rock falls 2 meters straight down in one second.  On this planet, how much does the rock weigh?                                    (b)CM

    a. 1 N.             d. 5 N.
    b. 4 N.             e. none of these.
    c. 4.9 N.

# 4

# Newton's Laws of Motion

1. A sheet of paper can be withdrawn from under a container of milk without toppling it if the paper is jerked quickly. The reason this can be done is that    (d)A

    a. the milk carton has no acceleration.
    b. there is an action-reaction pair operating.
    c. the gravitational field pulls on the milk carton.
    d. the milk carton has inertia.
    e. none of these.

2. An object maintains its state of motion because it has    (a)A

    a. mass                d. acceleration.
    b. momentum.           e. all of these.
    c. speed.

3. According to Newton's law of inertia, the result of walking much of the day is that we are actually a little bit shorter in the    (b)A

    a. morning.            c. ...no difference, really.
    b. evening.

4. Inertia is a measure of an object's    (c)A

    a. weight.             d. gravity.
    b. force.              e. center of mass.
    c. mass.

5. Your weight is a measure of your    (c)A
    a. momentum.
    b. energy.
    c. gravitational attraction to the Earth.
    d. rotational equilibrium.
    e. all of these.

6. One object has twice as much mass as another object.    The    (a)A
   first object also has twice as much

    a. inertia.            d. energy.
    b. velocity.           e. all of these.
    c. gravitational acceleration.

7. A 10 kg brick and a 1 kg book are dropped in a vacuum. The force of gravity on the 10 kg brick is    (b)A

    a. the same as the force on the 1 kg book.
    b. 10 times as much as the force on the 1 kg
       book.
    c. zero.

8. Compared to its weight on Earth, a 10 kg object on the Moon will weigh  (a)A

    a. less.              c. the same amount.
    b. more.

9. Compared to its mass on Earth, the mass of an object on the Moon is  (c)A

    a. less.              c. the same.
    b. more.

10. When the net force acting on an object doubles, the acceleration  (b)A

    a. quadruples.       d. halves.
    b. doubles.         e. none of these.
    c. remains the same.

11. Suppose an object's mass is decreasing.  If a constant force is applied to the object, the acceleration  (b)A

    a. decreases.       c. remains the same.
    b. increases.

12. An object is pulled northward with a force of 10 N and southward with a force of 15 N.  The magnitude of the net force on the object is  (b)A

    a. 0 N.         d. 15 N.
    b. 5 N.         e. none of these.
    c. 10 N.

13. Strange as it may seem, it is just as hard to push a car on the Moon as it is to push the same car on Earth.  This is because  (a)A

    a. the mass of the car is independent of gravity.
    b. the weight of the car is independent of gravity.
    c. ... Nonsense! A car is much more easily pushed on the Moon than on the Earth.

14. An object is propelled along a straight-line path by a force.  If the force were doubled, its acceleration would  (b)A

    a. quadruple       d. half.
    b. double.        e. none of these.
    c. stay the same.

15. An object is propelled along a straight-line path in space by a force.  If the mass of the object doubles its acceleration  (d)A

    a. quadruples.       d. halves.
    b. doubles.         e. none of these.
    c. stays the same.

16. The force of friction on a sliding object is 10 Newtons.  The force needed to maintain a constant velocity is  (c)A

    a. more than 10 N.    c. 10 N.
    b. less than 10 N.

17. A 10 N falling object encounters 4 N of air resistance.     (c)A
    The magnitude of the net force on the object is

        a. 0 N.                 c. 10 N.
        b. 4 N.                 d. none of these.
        c. 6 N.

18. A 10 N falling object encounters 10 N of air resistance.    (a)A
    The magnitude of the net force on the object is

        a. 0 N.                 c. 10 N.
        b. 4 N.                 d. none of these.
        c. 6 N.

19. An apple weighs 1 N.  When held above your head, the net    (a)A
    force on the apple is

        a. 0 N.                 d. 9.8 N.
        b. 0.1 N.               e. none of these.
        c. 1 N.

20. An apple weighs 1 N. The net force on the apple when it is   (c)A
    in free fall is

        a. 0 N.                 d. 9.8 N.
        b. 0.1 N.               e. none of these.
        c. 1 N.

21. A car has a mass of 1000 kg and accelerates at 2 meters per  (d)AM
    second squared.  What is the magnitude of the force acting
    on the car?

        a. 500 N.               d. 2000 N.
        b. 1000 N.              e. none of these.
        c. 1500 N.

22. A tow-truck exerts a force on 3000 N on a car, accelerating  (c)AM
    it at 2 meters per second squared.  What is the mass of the
    car?

        a. 500 kg.              d. 3000 kg.
        b. 1000 kg.             e. none of these.
        c. 1500 kg.

23. A girl pulls on a 10 kg wagon with a constant force of       (b)AM
    30 N.  What is the wagon's acceleration?

        a. 0.3 meters per second squared.
        b. 3.0 meters per second squared.
        c. 10 meters per second squared.
        d. 30 meters per second squared.
        e. 300 meters per second squared.

24. An object has a constant mass.  A constant force on the      (b)B
    object produces constant

        a. velocity.            c. both of the above
        b. acceleration.        d. none of the above.

25. A force of 1 N accelerates a mass of 1 kg at the rate of     (c)B
    1 meter per second squared.  The acceleration of a mass of
    2 kg acted upon by a force of 2 N is

        a. half as much.        c. the same.
        b. twice as much.       d. none of these.

26. A bag of groceries has a mass of 10 kilograms and a weight        (c)B
    of about

        a. 1 N.                    d. 1000 N.
        b. 10 N.                   e. none of these.
        c. 100 N.


27. The mass of a dog that weighs 100 N is about                      (b)B

        a. 1 kg.                   d. 1000 kg.
        b. 10 kg.                  e. none of these.
        c. 100 kg.


28. The force required to maintain an object at a constant            (a)B
    speed in free space is equal to

        a. zero.
        b. the mass of the object.
        c. the weight of the object.
        d. the force required to stop it.
        e. none of these.


29. An object following a straight-line path at constant speed        (b)B

        a. has a net force acting upon it in the direction of
           motion.
        b. has zero acceleration.
        c. must be moving in a vacuum.
        d. has no forces acting on it.
        e. none of the above.


30. A man weighing 800 N stands on two bathroom scales so that        (b)B
    his weight is distributed evenly over the scales.  The
    reading on each scale is

        a. 200 N.                  d. 1600 N.
        b. 400 N.                  e. none of these
        c. 800 N.


31. Neglecting friction, a Cadillac and Volkswagon start roll-        (c)B
    ing down a hill together.  The heavier Cadillac will get to
    the bottom

        a. before the Volkswagon.
        b. after the Volkswagon.
        c. the same time as the Volkswagon.


32. A hockey puck is set in motion across a frozen pond. If ice       (a)B
    friction and air resistance are neglected, the force
    required to keep the puck sliding at constant velocity is

        a. zero newtons.
        b. equal to the weight of the puck.
        c. the weight of the puck divided by the mass of the puck.
        d. the mass of the puck multiplied by 9.8 meters per second
           squared.
        e. none of these.


33. When a woman stands with two feet on a scale, the scale           (c)B
    reads 500 N.  When she lifts one foot, the scale reads

        a. less than 500 N.        c. 500 N.
        b. more than 500 N.

34. A push on a one kilogram brick accelerates the brick.    (b)B
    Neglecting friction, to equally accelerate a 10 kilogram
    brick, one would have to push

    a. with just as much force.
    b. with 10 times as much force.
    c. with 100 times as much force.
    d. with one tenth the amount of force.
    e. none of these.


35. A 10 N block and a 1 N block lie on a horizontal friction-    (b)B
    less table.  To push them with equal acceleration, we would
    have to push with

    a. equal forces on each block.
    b. ten times as much force on the heavier block.
    c. ten squared or 100 times as much force on the heavier
       block.
    d. one tenth as much force on the heavier block.
    e. none of these.


36. A rocket becomes progressively easier to accelerate as it    (d)B
    travels in space away from the Earth because

    a. gravitation becomes weaker with increased distance
       from the Earth.
    b. the net force on the rocket increases as burning of
       fuel progresses.
    c. the mass of the rocket decreases as fuel is burned.
    d. all of the above.
    e. none of the above.


37. A rock is thrown vertically into the air.  At the very top    (c)B
    of its trajectory the net force on it is

    a. less than its weight.
    b. more than its weight.
    c. its weight.


38. A heavy pile driver starting from rest falls on a pile with    (d)B
    a force that depends on

    a. the original height of the driver.
    b. the original potential energy of the driver.
    c. the distance the driver falls.
    d. all of the above.
    e. none of the above.


39. A truck is moving at constant speed.  Inside the storage    (a)B
    compartment, a rock is dropped from the midpoint of the
    ceiling and strikes the floor below.  The rock hits the
    floor

    a. exactly below the midpoint of the ceiling.
    b. ahead of the midpoint of the ceiling.
    c. behind the midpoint of the ceiling.
    d. more information is needed to solve this problem.
    e. none of these.


40. A block is dragged without acceleration in a straight line    (c)B
    path across a level surface by a force of 6 N.  What is the
    frictional force between the block and the surface?

    a. less than 6 N.        c. 6 N.
    b. more than 6 N.        d. need more information to say.

41. Suppose a particle is accelerated through space by a 10 N          (c)BM
    force.  Suddenly the particle encounters a second force of
    10 N in the opposite direction of the first force.  The
    particle

        a. is brought to a rapid halt.
        b. decelerates gradually to a halt.
        c. continues at the speed it had when it encountered
           the second force.
        d. theoretically accelerates to speeds approaching the
           speed of light in the absence of a net force.
        e. none of these.


42. A 747 jumbo jet has a mass of 30,000 kg. The thrust for each       (d)BM
    of four engines is 30,000 N.  What is the jet's acceleration
    when taking off?

        a. 0.25 meters per second squared
        b. 1 meter per second squared.
        c. 2.5 per second squared.
        d. 4 meters per second squared.
        e. none of these.


43. In which case would you have the largest mass of gold? If          (a)C
    your chunk of gold weighed 1 N on the

        a. Moon.                c. planet Jupiter.
        b. Earth.


44. An object weighs 30 N on Earth.  A second object weighs            (b)C
    30 N on the Moon.  Which has the greater mass?

        a. the one on Earth.
        b. the one on the Moon.
        c. they have the same mass.
        d. not enough information to say.


45. As a bird sits on the limb of a tree, both the bird and the tree are  (c)C
    moving at about 0.3 mi/s due to the Earth's rotation about
    its axis.  Several hundred years ago, people philosophized
    that when the bird leaves the tree and flies into the air,
    the tree would go whizzing past the bird at 0.3 mi/s.  The
    reason this does not happen has to do with Newton's

        a. first law.           d. law of gravitation.
        b. second law.          e. none of these.
        c. third law.


46. A skydiver of mass 100 kilograms experiences air resistance        (d)CM
    of 500 Newtons.  What is the diver's acceleration?

        a. 0.2g                 d. 0.5g
        b. 0.3g                 e. more than 0.5g
        c. 0.4g.


47. An object released from rest on another planet requires            (c)CM
    one second to fall a distance of 6 meters.  What is the
    acceleration due to gravity on this planet?

        a. 3 meters per second squared.
        b. 6 meters per second squared.
        c. 12 meters per second squared.
        d. 15 meters per second squared.
        e. none of these.

48. A car traveling at 22 m/s hits a tree.  The collision time    (b)CM
    is 0.1 second.  What is the deceleration of the car?

    a. 110 meters per second squared.
    b. 220 meters per second squared.
    c. 800 meters per second squared.
    d. 880 meters per second squared.
    e. none of these.

49. A 10-kilogram block with an initial velocity of 10 m/s        (d)CM
    slides 10 meters across a horizontal surface and comes to
    rest.  It takes the block 2 seconds to stop.  The stopping
    force acting on the block is about

    a. 5 N.                    d. 50 N.
    b. 10 N.                   e. none of these.
    c. 25 N.

50. A 10-kilogram block is pushed with a horizontal force of      (a)CM
    20 N against a friction force of 10 N.  The block
    accelerates at

    a. 1 meter per second squared.
    b. 2 meters per second squared.
    c. 5 meters per second squared.
    d. 10 meters per second squared.
    e. none of these.

51. The deceleration of an automobile in a full skid on dry       (b)CM
    pavement is about 1/2 g.  If you are driving at 22 m/s
    and slam on your brakes, the time required to a stop is
    about

    a. 3.5 seconds.            d. 6.5 seconds.
    b. 4.5 seconds.            e. more than 6.5 seconds.
    c. 5.5 seconds.

52. A 1000-kg automobile enters a freeway on-ramp at  20 m/s       (c)CM
    and accelerates uniformly up to 40 m/s in a time of 10
    seconds.  How far does the automobile travel during that
    time?

    a. 100 m.                  d. 400 m.
    b. 200 m.                  e. none of these.
    c. 300 m.

53. A 2000-kg car experiences a braking force of 10,000 N and     (c)CM
    skids to a stop in 6 seconds.  The speed of the car just
    before the brakes were applied was

    a. 1.2 m/s.                d. 45 m/s.
    b. 15 m/s.                 e. none of these.
    c. 30 m/s.

54. An astronaut on another planet drops a 1-kg rock from rest    (e)CM
    and finds that it falls a vertical distance of 4 meters in
    one second.  On this planet, the rock has a weight of

    a. 1 N.                    d. 5 N.
    b. 4 N.                    e. none of these.
    c. 4.9 N.

Chapter 4c: Newton's Third Law

1. An archer shoots an arrow.  Consider the action force to be                    (e)A
   the bowstring against the arrow.  The reaction to this force
   is the

   a. weight of the arrow.
   b. air resistance against the bow.
   c. friction of the ground against the archer's feet.
   d. grip of the archer's hand on the bow.
   e. arrow's push against the bowstring.


2. A player catches a ball.  Consider the action force to be                      (b)A
   the impact of the ball against the player's glove.  What
   is the reaction to this force?

   a. the player's grip on the glove.
   b. the force the glove exerts on the ball.
   c. friction of the ground against the player's shoes.
   d. the muscular effort in the player's arms.
   e. none of these.


3. A player hits a ball with a bat.  The action force is the                      (c)A
   impact of the bat against the ball.  What is the reaction
   to this force?

   a. air resistance on the ball.
   b. the weight of the ball.
   c. the force of the ball against the bat.
   d. the grip of the player's hand against the ball.
   e. none of these.


4. A baseball player bats a ball with a force of 1000 N.  The                     (c)A
   ball exerts a reaction force against the bat of

   a. less than 1000 N.      c. 1000 N.
   b. more than 1000 N.


5. As a ball falls, the action force is the pull of the                          (c)B
   earth's mass on the ball.  What is the reaction to this
   force?

   a. air resistance acting against the ball.
   b. the acceleration of the ball.
   c. the pull of the ball's mass on the earth.
   d. ... non-existent in this case.
   e. none of these.


6. A person is attracted towards the center of the earth by                      (c)B
   a 500 N. gravitational force.  The force that the earth is
   attracted toward the person is

   a. very very small.       c. 500 N.
   b. very very large.

7. The attraction of a person's body toward the earth is called     (c)B
   weight. The reaction to this force is

    a. the person's body pushing against the earth's surface.
    b. the earth's surface pushing against the person's body.
    c. the person's body pulling on the earth.
    d. none of these.

8. As a high-diver dives towards the earth, the attraction of     (e)B
   the Earth on the diver's body pulls the diver down. What is
   the reaction to this force?

    a. air resistance the diver encounters while falling.
    b. water resistance that will soon act upward on the
       diver.
    c. the attraction to the planets, stars, and every
       particle in the universe.
    d. all of the above.
    e. none of the above.

9. A Mack truck and a Volkswagen traveling at the same speed     (a)B
   have a head-on collision. The vehicle to undergo the
   greatest change in velocity will be the

    a. Volkswagen.        c. both the same.
    b. Mack truck.

10. A car traveling at 100 km/hr strikes a hapless bug and     (c)B
    splatters it. The force of impact is greater on the

    a. bug.        c. ...is the same for both.
    b. car.

11. The force exerted on the tires of a car to directly     (d)B
    accelerate it along a road is exerted by the

    a. engine.        d. road.
    b. tires.        e. none of these.
    c. air.

12. Two people pull on a rope in a tug-of-war. Each pulls with     (b)B
    400 N of force. What is the tension in the rope?

    a. zero.        d. 800 N.
    b. 400 N.       e. none of these.
    c. 600 N.

13. A Mack truck and a Volkswagen traveling at the same speed     (c)B
    collide head-on. The impact force is greatest on the

    a. Volkswagen.        c. ...is the same
    b. Mack truck.          for both.

14. A horse exerts 500 N of force on a heavy wagon. The     (b)C
    wagon pulls back on the horse with an equal force. The
    wagon still accelerates because

    a. these forces are not an action-reaction pair.
    b. nevertheless there is still an unbalanced force on
       the wagon.
    c. the horse pulls on the wagon a brief time before
       the wagon reacts.
    d. ...the wagon does not accelerate because these
       forces are equal and opposite.
    e. none of these.

15. Two people, one twice as massive as the other, attempt a    (b)CM
    tug-of-war with 12 meters of rope on frictionless ice.
    After a brief time, they meet.  The heavier person will
    have slid a distance of

            a. 3 m.                 d. 6 m.
            b. 4 m.                 e. not enough information
            c. 5 m.                    given to say.


16. A rifle of mass 2 kg is suspended by strings.  The rifle    (d)CM
    fires a bullet of mass 1/100 kg at a speed of 200 m/s.
    What is the recoil velocity of the rifle?

            a. 0.001 m/s.           d. 1 m/s.
            b. 0.01 m/s.            e. none of these.
            c. 0.1 m/s.


- 51 -

# 5

# Momentum

Chapter 5: Momentum

1. Which of the following has the largest momentum?                    (b)A

    a. a tightrope walker crossing Niagara Falls.
    b. a pickup truck traveling down the highway.
    c. a Mack truck parked in a parking lot.
    d. the Science building at City college.
    e. a dog running down the street.

2. A freight train rolls along a track with considerable           (b)A
   momentum.  If it rolls at the same speed but has twice as
   much mass, its momentum is

    a. zero.                  c. quadrupled.
    b. doubled.               d. unchanged.

3. A moving object on which no forces are acting will continue       (c)A
   to move with constant

    a. acceleration.          d. all of the above
    b. impulse.               e. none of the above.
    c. momentum.

4. A rifle recoils from firing a bullet.  The speed of the          (c)A
   rifle's recoil is small because the

    a. force against the rifle is relatively small.
    b. speed is mainly concentrated in the bullet.
    c. rifle has lots of mass.
    d. momentum of the rifle is unchanged.
    e. none of these.

5. Suppose a gun were made of a strong but very light material.      (a)A
   Suppose also that the bullet is more massive than the gun
   itself. For such a weapon

    a. the target would be a safer place than where the shooter
       is.
    b. recoil problems would be lessened.
    c. conservation of energy would not hold.
    d. conservation of momentum would not hold.
    e. both conservation of energy and momentum would not hold.

6. Two objects, A and B, have the same size and shape, but A         (c)A
   is twice as heavy as B.  When they are dropped simultaneously from a tower, they reach the ground at the same time,
   but A has a higher

    a. speed.                 d. all of the above.
    b. acceleration.          e. none of the above.
    c. momentum.

7. A heavy truck and a small truck roll down a hill.          (c)A
   Neglecting friction, at the bottom of the hill, the heavy
   truck will have greater

        a. speed.              d. all of the above.
        b. acceleration.       e. none of the above.
        c. momentum.


8. In order to catch a ball, a baseball player moves his or   (c)A
   her hand backward in the direction of the ball's motion.
   Doing this reduces the force of impact on the players hand
   principally because

        a. the resultant velocity of impact is lessened.
        b. the momentum of impact is reduced.
        c. the time of impact is increased.
        d. the time of impact is decreased.
        e. none of these.


9. A car traveling along the highway needs a certain amount of (d)A
   force exerted on it to stop.  More stopping force may be
   required when the car has

        a. more mass.          d. all of the above.
        b. more momentum.      e. none of the above.
        c. less stopping distance.


10. A gun fires a bullet.  The speed of the bullet will be the (d)A
    same as the speed of the recoiling gun

        a. because momentum is conserved.
        b. because velocity is conserved.
        c. because both velocity and momentum are conserved.
        d. if the mass of the bullet equals the mass of the gun.
        e. none of these.


11. Padded dashboards in cars are safer in an accident than    (d)A
    nonpadded ones because they

        a. increase the impact time.
        b. decrease an occupant's impulse.
        c. decrease the impact force.
        d. all of the above.
        e. none of the above.


12. The force on an apple hitting the ground depends upon       (e)A

        a. air resistance on the apple as it falls.
        b. the speed of the apple just before it hits.
        c. the time of impact with the ground.
        d. whether or not the apple bounces.
        e. all of these.


13. When you jump from an elevated position you usually bend    (c)A
    your knees upon reaching the ground.  By doing this, the
    time of the impact is about 10 times less what is would be
    in a stiff-legged landing.  In this way the average force
    your body experiences is reduced by

        a. less than 10 times.   c. about 10 times.
        b. more than 10 times.


14. A 2 kg ball is thrown at 3 m/s.  What is the ball's         (c)AM
    momentum?

        a. 2 kg*m/s.             d. 9 kg*m/s.
        b. 3 kg*m/s.             e. none of these.
        c. 6 kg*m/s.

15. A 4 kg ball has a momentum of 12 kg*m/s.  What is the          (a)AM
    ball's speed?

        a. 3 m/s.              d. 48 m/s.
        b. 4 m/s.              e. none of these.
        c. 12 m/s.

16. A ball is moving at 4 m/s and has a momentum of 48 kg*m/s.     (b)AM
    What is the ball's mass?

        a. 4 kg.              d. 192 kg.
        b. 12 kg.            e. none of these.
        c. 48 kg.

17. Recoil is noticeable if we throw a heavy ball while           (a)B
    standing on roller skates.  If instead we go through the
    motions of throwing the ball but hold onto it, our net
    recoil will be.

        a. zero.              c. small, but noticeable.
        b. the same as before.

18. A heavy truck and a small car rolling down a hill at the      (a)B
    same speed are forced to stop in the same amount of time.
    Compared to the force that stops the car, the force needed
    to stop the truck is

        a. greater.          c. the same.
        b. smaller.

19. A 1 N apple falls to the ground.  The apple hits the ground   (e)B
    with an impact force of about

        a. 1 N.              d. 9.8 N.
        b. 2 N.              e. not enough information
        c. 4 N.                 given to say.

20. A karate expert executes a swift blow and severs a cement     (d)B
    block with her or his bare hand.  The

        a. impulse on both the block and the expert's hand
           have the same magnitude.
        b. force on both the block and the expert's hand
           have the same magnitude.
        c. time of impact on both the block and the expert's
           hand is the same.
        d. all of the above.
        e. none of the above.

21. A piece of putty moving with 1 unit of momentum strikes and   (c)B
    sticks to a heavy bowling ball that is initially at rest.
    After the putty sticks to the ball, both are set in motion
    with a combined momentum that is

        a. less than 1 unit.    c. 1 unit.
        b. more than 1 unit.    d. not enough information.

22. A 1 kg chunk of putty moving at a 1 m/s collides and sticks   (b)B
    to a 5 kg bowling ball that is initially at rest. The
    bowling ball with its putty host is then set in motion with
    a momentum of

        a. 0 kg m/s.         d. 5 kg m/s.
        b. 1 kg m/s.         e. more than 5 kg m/s.
        c. 2 kg m/s.

23. The force that accelerates a rocket in outer space is        (d)B
    exerted on the rocket by the

        a. rocket's engine.      d. exhaust gases.
        b. rocket's wings.       e. none of these.
        c. atmospheric pressure.

24. If the polar icecaps melted, the resulting water would        (a)B
    spread over the entire Earth.  This new mass distribution
    would tend to make the length of a day

        a. longer.               c. stay the same.
        b. shorter.

25. Two billiard balls having the same mass roll toward each       (a)B
    other, each moving at the same speed.  What is the combined
    momentum of the two balls?

        a. 0 kg*m/s.             c. more information need to
        b. 10 kg*m/s.               determine.

26. A 1-kg chunk of putty moving at 1 m/s collides with and        (c)BM
    sticks to a 5-kg bowling ball that is initially at rest.
    The bowling ball with its putty host is then set in motion
    with a speed of

        a. 1/4 m/s.              d. none of these.
        b. 1/5 m/s.              e. not enough information given.
        c. 1/6 m/s.

27. Momentum is transferred to the ground when an apple falls      (b)C
    on it.  The momentum absorbed by the ground is

        a. negligible compared to the momentum of the apple.
        b. greater than that of the apple only if the apple
           bounces.
        c. greater than that of the apple in all cases.
        d. none of these.

28. According to the impulse-momentum equation Ft = change in      (c)C
    (mv), a person will suffer less injury falling on a wooden
    floor which "gives" than on a more rigid cement floor. The
    "F" in the above equation stands for the force exerted on
    the

        a. person.              c. both of the above.
        b. floor.               d. none of the above.

29. If all people, animals, trains and trucks all over the         (b)C
    world began to walk or run towards the east, then the

        a. Earth would spin a bit faster.
        b. Earth would spin a bit slower.
        c. the Earth's spin would not be affected at all.

30. You're driving down the highway and a bug spatters into        (c)C
    your windshield.  Which undergoes the greater change in
    momentum?

        a. the bug.             c. both the same.
        b. your car.

31. Suppose an astronaut in outer space wishes to play a          (b)C
    solitary "throw, bounce, and catch" game by tossing a ball
    against a very massive and perfectly elastic concrete wall.
    If the ball is as massive as the astronaut

        a. the astronaut will catch one bounce only.
        b. the astronaut will never catch the first bounce.
        c. the astronaut's time between catches will decrease as
           the game progresses.
        d. none of these.

32. A golf ball moving forward with 1 unit of momentum strikes     (b)C
    and bounces backward off a heavy bowling ball that is
    initially at rest and free to move.  The bowling ball is
    set in motion with a momentum of

        a. less than 1 unit.     c. 1 unit.
        b. more than 1 unit.     d. not enough information.

# 6

## Energy

Chapter 6a: Work and Energy

1. An object is raised above the ground gaining a certain        (b)A
   amount of potential energy.  If the same object is raised
   twice as high it gains

       a. four times as much potential energy.
       b. twice as much potential energy.
       c. neither of these.

2. When an object is lifted 10 meters, it gains a certain        (c)A
   amount of potential energy. If the same object is lifted 20
   meters, its potential energy is

       a. less.          d. four times as much.
       b. the same.      e. more than 4 times
       c. twice as much.     as much.

3. A 1000 kg car and a 2000 kg car are hoisted the same        (c)A
   distance in a gas station. Raising the more massive car
   requires
       a. less work.       d. four times as much work.
       b. as much work.    e. more than 4 times as much
       c. twice as much work.   work.

4. An object that has kinetic energy must be        (a)A

       a. moving.        d. at rest.
       b. falling.      e. none of these.
       c. at an elevated position.

5. An object that has potential energy has this energy because        (d)A
   of its

       a. speed.        d. location.
       b. acceleration.    e. none of these.
       c. momentum.

6. Bullets are fired from an airplane in the forward direction        (a)A
   of motion.  The momentum of the airplane will be

       a. decreased.      c. increased.
       b. unchanged.

7. A workwoman can lift barrels a vertical distance of 1 meter        (a)A
   or can roll them up a 2 meter-long ramp to the same
   elevation.  If she uses the ramp, the applied force
   required is

       a. half as much.    c. the same.
       b. twice as much.

8. An arrow is drawn so that it has 40 J of potential energy.   (c)A
When fired, the arrow will have a kinetic energy of

    a. less than 40 J.        c. 40 J.
    b. more than 40 J.


9. A job is done slowly, and an identical job is done            (b)A
quickly.  Both jobs require the same amount of work, but
different amounts of

    a. energy.                c. both of the above.
    b. power.                 d. none of the above.


10. Which requires more work:  lifting a 50 kg sack vertically   (c)AM
2 meters or lifting a 25 kg sack vertically 4 meters?

    a. lifting the 50 kg sack.
    b. lifting the 25 kg sack.
    c. both require the same amount of work.


11. A 50 kg sack is lifted 2 meters in the same time as a 25 kg  (c)AM
sack is lifted 4 meters.  The power expended in raising the
50 kg sack compared to the power used to lift the 25 kg
sack is

    a. twice as much.         c. the same.
    b. half as much.


12. A TV set is pushed a distance of 2 m with a force of 20 N    (d)AM
that is in the same direction as the set moves.  How much
work is done on the set?

    a. 2 J.                   d. 40 J.
    b. 10 J.                  e. 80 J.
    c. 20 J.


13. It takes 40 J to push a large box 4 m across a floor.        (b)AM
Assuming the push is in the same direction as the move,
what is the magnitude of the force on the box?

    a. 4 N.                   d. 160 N.
    b. 10 N.                  e. none of these.
    c. 40 N.


14. A man lifts a box that weighs 20 N a distance of 1 m.  How   (a)AM
much work does the man do on the box?

    a. 20 J.                  d. 80 J.
    b. 40 J.                  e. none of these.
    c. 60 J.


15. A 2 kg mass is held 4 m above the ground.  What is the       (d)AM
approximate potential energy of the mass with respect to
the ground?

    a. 20 J.                  d. 80 J.
    b. 40 J.                  e. none of these.
    c. 60 J.


16. A 2 kg mass has 40 J of potential energy with respect to     (b)AM
the ground.  Approximately how far is it located above the
ground?

    a. 1 m.                   d. 4 m.
    b. 2 m.                   e. none of these.
    c. 3 m.

17. Using 1000 J of work, a toy elevator is raised from the     (b)AM
    ground floor to the second floor in 20 seconds.  How much
    power does the elevator use?

    a. 20 W.          d. 1000 W.
    b. 50 W.          e. 20000 W.
    c. 100 W.

18. One end of a long uniform log is raised to shoulder level.    (b)B
    Another identical log is raised at its center to the same
    level.  Raising the second log requires about

    a. the same amount of work.
    b. twice as much work.
    c. more than twice as much work.

19. Two identical arrows, one with twice the kinetic energy of    (b)B
    the other, are fired into a hay bale.  The faster arrow
    will penetrate

    a. the same distance as the slower arrow.
    b. twice as far as the slower arrow.
    c. four times as far as the slower arrow.
    d. more than four times as far as the slower arrow.
    e. none of these.

20. The kinetic energy of a satellite in elliptical orbit about   (a)B
    the earth is greatest at

    a. perigee (nearest point).
    b. apogee (farthest point).
    c. ...is the same everywhere.

21. A ball is projected into the air with 100 J of kinetic        (a)B
    energy which is transformed to gravitational potential
    energy at the top of its trajectory.  When it returns to
    its original level after encountering air resistance, its
    kinetic energy is

    a. less than 100 J.    c. 100 J.
    b. more than 100 J.    d. not enough information given.

22. Strictly speaking, a car will burn more gasoline if the car   (d)B
    radio is turned on.  This statement is

    a. totally false.
    b. true only if the car's engine is running.
    c. true only if the car's engine is stopped.
    d. always true.
    e. none of these.

23. A machine puts out 100 Watts of power for every 1000 Watts    (a)B
    put into it.  The efficiency of the machine is

    a. 10 %.          d. 110 %.
    b. 50 %.          e. none of these.
    c. 90 %.

24. An ungloved fist will do more damage to a jaw than a gloved    (b)B
    fist.  The reason for this is that the ungloved fist

    a. delivers a larger impulse to the jaw.
    b. exerts a larger force on the jaw.
    c. has less air resistance on it.
    d. none of these.

25. A woman lifts a box from the floor.  She then moves with             (a)B
    constant speed to the other side of the room, where she
    puts the box down.  How much work does she do on the box
    while walking across the floor at constant speed?

        a. zero J.               c. more information need to
        b. more than zero J.        determine.

26. A car moving at 50 km/hr skids 20 m with locked brakes. How          (e)BM
    far will the car skid with locked brakes if it is traveling
    at 150 km/hr?

        a. 20 m.                 d. 120 m.
        b. 60 m.                 e. 180 m.
        c. 90 m.

27. Which has greater kinetic energy, a car traveling at 30              (b)BM
    km/hr or a half-as-massive car traveling at 60 km/hr?

        a. the 30 km/hr car.
        b. the 60 km/hr car.
        c. both have the same kinetic energy.

28. A diver who weighs 500 N steps off a diving board that is           (d)BM
    10 m above the water.  The diver hits the water with
    kinetic energy of

        a. 10 J.                 d. 5000 J.
        b. 500 J.                e. more than 5000 J.
        c. 510 J.

29. Consider a hydraulic press.  When the input piston is               (c)BM
    depressed 20 cm, the output piston is observed to move
    1 cm.  On the same press, an input force of 1 N can raise

        a. 1 N.                  c. 20 N.
        b. 10 N.                 d. more than 20 N.

30. A 2500 N pile driver ram falls 10 m and drives a post 0.1 m         (c)BM
    into the ground.  The average impact force on the ram is

        a. 2500 N.               c. 250,000 N.
        b. 25000 N.              d. 2,500,000 N.

31. A pulley system raises a 1000 N load with 100 N of input            (e)BM
    force.  The efficiency of the system is

        a. 10 %.                 d. 1000 %.
        b. 90 %.                 e. not enough information
        c. 100 %.                   given.

32. A jack system will increase the potential energy of a heavy         (c)BM
    load by 1000 J with a work input of 2000 J.  The efficiency
    of the jack system is

        a. 10 %.                 d. 80 %.
        b. 20 %.                 e. not enough information given.
        c. 50 %.

33. If a power plant is 30 % efficient, and the transmission            (d)BM
    system  that delivers power to comsumers is 60 % efficient,
    then the overall efficiency is

        a. 90 %.                 d. 18 %.
        b. 60 %.                 e. none of these.
        c. 30 %.

34. Two identical arrows, one with twice the speed of the other, are fired into a hay bale. The faster arrow will penetrate

    a. the same distance as the slower arrow.
    b. twice as far as the slower arrow.
    c. four times as far as the slower arrow.
    d. more than four times as far as the slower arrow.
    e. none of these.

35. A person on the edge of a roof throws a ball downward. It strikes the ground with 100 J of kinetic energy. The person throws another identical ball upward with the same initial speed, and this too falls to the ground. Neglecting air resitance, the second ball hits the ground with a kinetic energy of

    a. 100 J.                 d. more than 200 J.
    b. 200 J.                 e. none of these.
    c. less than 100 J.

36. A 10 N object moves at 1 m/s. Its kinetic energy is

    a. 0.5 J.                 c. 10 J.
    b. 1 J.                   d. more than 10 J.

37. A ball dropped onto a steel plate loses 10 % of its kinetic energy with each bounce. The number of bounces the ball makes before it fails to rise to half its initial height is

    a. 5.                     d. more than 10.
    b. 7.                     e. not enough information
    c. 10.                       given.

38. How many Joules of energy are in one kilowatt-hour?

    a. 1 J.                   d. 3.6 megajoules.
    b. 60 J.                  e. none of these.
    c. 60 kilojoules.

39. A car's engines is 20 % efficient. When cruising, the car encounters an average retarding force of 1000 N. If the energy content of gasoline is 40 megajoules per liter, how many kilometers per liter does the car get?

    a. 14 km/l.               d. 8 km/l.
    b. 12 km/l.               e. none of these.
    c. 10 km/l.

40. Suppose a miracle car has a 100 % efficient engine and burns fuel that has a 40 megajoules per liter energy content. If the air drag and overall frictional forces on this car traveling at highway speeds is 2000 N, what is the overall limit in distance per liter it could be driven on the highway?

    a. 15 km.                 d. more than 25 km.
    b. 20 km.                 e. not enough information.
    c. 25 km.

41. An insurance office uses 1 KW of lighting. A liter of oil burned at the power plant gets 25 % energy conversion. Assuming oil contains 36 megajoules of energy per liter, about how long will one liter light the office?

    a. 2.5 hrs.               d. 10 hrs.
    b. 5 hrs.                 e. more than 10 hrs.
    c. 7.5 hrs.

Chapter 6b: Momentum and Energy

1. A feather and a coin dropped in a vacuum fall with equal          (c)A

    a. forces.              d. kinetic energies.
    b. momenta.             e. none of these.
    c. accelerations.


2. When a rifle is fired it recoils so both the bullet and          (a)A
   rifle are set in motion.  The gun and bullet acquire equal

    a. but opposite amounts of momentum.
    b. amounts of kinetic energy.
    c. both of the above.
    d. none of the above.


3. A moving object has                                              (e)A

    a. speed.               d. energy.
    b. velocity.            e. all of these.
    c. momentum.


4. An object at rest may have                                       (c)A

    a. speed.               d. momentum.
    b. velocity.            e. none of these.
    c. energy.


5. A heavy and a light object released from the same height in      (e)A
   a vacuum have equal

    a. weights.             d. all of the above.
    b. momenta.             e. none of the above.
    c. energies.


6. What does an object have when moving that it doesn't have        (a)A
   when at rest?

    a. momentum.            d. inertia.
    b. energy.              e. none of these.
    c. mass.


7. If an object has kinetic energy, then it also must have          (b)A

    a. impulse.             d. force.
    b. momentum.            e. none of these.
    c. acceleration.


8. If the speed of a moving object doubles, then what else          (a)A
   doubles?

    a. momentum.            d. all of the above.
    b. kinetic energy.      e. none of the above.
    c. acceleration.

9. A feather and a coin are dropped in the air.  Each falls         (e)A
   with equal

      a. momenta.            d. all of the above.
      b. kinetic energies.    e. none of the above.
      c. potential energies.


10. Two pool balls, each moving at 2 m/s, roll toward each         (b)B
    other and collide.  Suppose after bouncing apart, each
    moves at 4 m/s.  This collision violates conservation of

      a. momentum.        c. both of the above.
      b. kinetic energy.    d. none of the above.


11. When a bullet is fired from a rifle, the force on the rifle    (a)B
    is equal to the force on the bullet.  However, the energy
    of the bullet is greater than the energy of the recoiling
    rifle because the

      a. force on the bullet acts for a longer distance.
      b. bullet's momentum is greater than that of the
         rifle.
      c. force on the bullet acts for a longer time.
      d. the impulse of the bullet is more.
      e. none of these.


12. An open freight car rolls friction free along a horizontal     (b)C
    track in a pouring rain that falls vertically.  As water
    accumulates in the car, its speed

      a. increases.       c. doesn't change.
      b. decreases.


13. A car has a head-on collision with another car of the same     (c)C
    momentum.  An identical car driving with the same speed as
    the first car runs into an enormously massive wall.  The
    greatest impact force will occur on the car that is in the
    collision with the

      a. approaching car.   c. ...both impact forces
      b. the wall.         will be the same.


14. A golf ball is thrown at and bounces away from a massive       (a)C
    bowling ball that is initially at rest.  After the
    collision, the bowling ball has more

      a. momentum, but less kinetic energy than the golf ball.
      b. kinetic energy, but less momentum than the golf ball.
      c. momentum and more kinetic energy than the golf ball.
      d. ... not enough information is given to say.


15. A popular swinging-balls apparatus consists of an aligned      (b)C
    row of identical elastic balls that are suspended by
    strings so they barely touch each other.  When two balls
    are lifted from one end and released, they strike the row
    and two balls pop out from the other end.  If instead one
    ball popped out with twice the velocity of the two, this
    would be violation of conservation of

      a. momentum.        c. both of the above.
      b. energy.          d. none of the above.

16. A piece of taffy slams into and sticks to another identical          (c)CM
    piece of taffy that is at rest.  The momentum of the two
    pieces stuck together after the collision is the same as it
    was before the collision, but this is not true of the
    kinetic energy, which is partly turned into heat.  What
    percentage of the kinetic energy is turned into heat?

        a. 0 %.                   d. 75 %.
        b. 25 %.                  e. not enough information
        c. 50 %.                     given.


17. Two identical freight cars roll without friction towards            (a)CM
    each other on a level track. One rolls at 2 m/s and the
    other rolls at 1 m/s.  After the cars collide, they couple
    and roll together with a speed of

        a. 0.5 m/s.               d. 1.0 m/s.
        b. 0.33 m/s.              e. none of these.
        c. 0.67 m/s.


18. A 1 kg ball dropped from a height of 2 m rebounds only 1.5 m        (e)CM
    after hitting the ground.  The amount of energy converted to
    heat is about

        a. 0.5 J.                 d. 2.0 J.
        b. 1.0 J.                 e. more than 2.0 J.
        c. 1.5 J.

# 7

# Rotational Motion

1. If the planet Jupiter underwent gravitational collapse, its    (b)A
   rate of rotation about its axis would

   a. decrease.             c. stay the same.
   b. increase.             d. more information needed.

2. The long heavy tail of a spider monkey enables the monkey    (d)A
   to easily vary its

   a. weight.               d. center of gravity.
   b. momentum.             e. none of these.
   c. inertia.

3. The center of mass of a human body is located at a point    (c)A

   a. which is fixed, but different for different people.
   b. that is always directly behind the belly button.
   c. which changes as a person bends over.
   d. none of these.

4. An object thrown into the air rotates about its    (c)A

   a. midpoint.             d. geometric center.
   b. edge.                 e. none of these.
   c. center of gravity.

5. A torque acting on an object tends to produce    (b)A

   a. equilibrium.          d. velocity.
   b. rotation.             e. a center of gravity.
   c. velcity.

6. "Centrifugal forces" are an apparent reality to observers    (d)A
   in a reference frame that is

   a. moving at constant velocity.
   b. an inertial reference frame.
   c. at rest.
   d. rotating.
   e. none of these.

7. A person weighs less at the equator than at the poles.  The    (a)A
   reason for this has to do with the

   a. spin of the Earth.
   b. tidal bulges.
   c. higher temperature at the equator, and the expansion
      of matter.
   d. influence of the Sun, Moon, and all the planets.
   e. none of these.

8. The rotational inertia of your leg is greater when your       (a)A
   leg is

    a. straight.        c. ...same either way.
    b. bent.

9. Horses that move with the fastest linear speed on a merry-    (b)A
   go-round are located

    a. near the center.    c. anywhere because they all
    b. near the outside.       move at the same speed.

10. Where does the pickup needle on a phonograph move with the   (b)A
    fastest speed?

    a. at the end of the record.
    b. at the beginning of the record.
    c. everywhere because it has the same speed anywhere
       on the record.

11. As the rotational speed of a space habitat increases, people's weight (a)A

    a. increases.        c. stays the same.
    b. decreases.

12. A huge rotating cloud of particles in space gravitate        (a)A
    together to form a more dense ball.  As the cloud shrinks
    in size it

    a. rotates faster.    c. rotates as the same
    b. rotates slower.       speed.

13. A car travels in a circle with constant speed.  The net      (b)B
    force on the car

    a. is directed forward, in the direction of travel.
    b. is directed towards the center of the curve.
    c. is zero because the car is not accelerating.
    d. none of these.

14. Two people are balanced on a see-saw.  If one person leans   (a)B
    toward the center of the see-saw, that person's end will

    a. rise.        c. stay at the same
    b. fall.        level.

15. To turn a stubborn screw, it is best to use a screwdriver    (a)B
    that has a

    a. wide handle.    c. yellow color.
    b. long handle.    d. none of these.

16. If the Earth rotated slower about its axis, your weight      (a)B
    would

    a. increase.       c. stay the same.
    b. decrease.

17. To weigh less in the northern hemisphere, you should go      (b)B

    a. north.       c. east.
    b. south.       d. west.

18. Suppose a certain carnival has a ferris wheel where the          (c)B
    seats are located on a radius, halfway between the center
    and outside rim.  Compared to an ordinary ferris wheel
    where the seats on are on the outside rim, your angular
    speed while riding on this ferris wheel would be

        a. less.                c. the same.
        b. more.

19. A coin and a ring roll down an incline starting at the same       (b)B
    time.  Which will reach the bottom first?

        a. the ring.            c. both reach the bottom
        b. the coin.               at the same time.

20. Which will reach the bottom of a hill sooner, an empty car        (c)C
    tire or the same tire mounted on a rim?

        a. the mounted tire provided the tire is more massive
           than the rim.
        b. the mounted tire provided the tire is less heavy
           than the rim.
        c. the mounted tire regardless of its weight.
        d. the empty tire provided it is as heavy as the rim.
        e. the empty tire regardless of its weight.

21. Suppose you put very large diameter tires on your car.            (b)C
    Then, your speedometer will read

        a. high.                c. the actual speed of the car.
        b. low.

22. A ring and a disk roll down a hill together.  Which reaches       (b)C
    the bottom first?

        a. the ring.
        b. the disk.
        c. depends on the masses.
        d. depends on the moments of inertia
        e. both reach the bottom at the same time.

23. A ring, a disk, and a solid sphere all having the same            (c)C
    radii  roll down a hill together.  Which reaches the bottom
    first?

        a. the ring.            d. ...they all reach
        b. the disk.               the bottom at the
        c. the sphere.             same time.

24. A ring, a disk, and a solid ball having equal masses roll         (c)C
    down a hill at the same time.  Which reaches the bottom
    first?

        a. the ring.            d. depends on what each is made of.
        b. the disk.            e. depends on the radii of each.
        c. the ball.

25. Neglecting air resistance, which will reach the bottom of         (a)C
    an incline sooner, an empty can, or the same can partially
    filled with beans?

        a. the filled can.      c. both reach the bottom
        b. the empty can.          at the same time.

26. A space habitat of the future is a cylinder rotating in        (a)C
    space about its long axis.  What is the relative
    gravitational field along the axis of the habitat?

        a. zero.                 d. one-half g.
        b. g.                    e. three-quarters g.
        c. one-quarter g.

27. A space habitat of the future may be a large cylinder          (c)C
    rotating about its long axis.  What is the gravitational
    field strength half-way between the axis and the outside
    edge where the field strength is g?

        a. zero.                 c. one-half g.
        b. g.                    d. three-quarters g.
        c. one-quarter g.

28. Consider a bowling alley inside a rotating space torus.        (a)C
    Compared to your bowling experience on Earth, you will
    notice the ball

        a. feels lighter when you throw it in the same direction
           as the bowling alley is moving and heavier when
           thrown in the opposite direction.
        b. feels heavier when you throw it in the same direction
           as the bowling alley is moving and lighter when
           thrown in the opposite direction.
        c. veers to the right when thrown in the same direction
           as the bowling alley is moving and to the left when
           thrown in the opposite direction.

29. A space habitat is designed so that the variation in g         (b)CM
    between a persons head and feet is less than 0.01 g.  If
    the person is 2 m tall, then the radius of the habitat is

        a. 20 m.                 c. 2000 m.
        b. 200 m.                d. more than 2000 m.

30. A swimming area in a rotating space habitat is located in a    (c)CM
    one-fourth g region.  If a diver can jump 1 m high in a one
    g region, how high can the same diver jump in the swimming
    area?

        a. 1 m.                  d. 16 m.
        b. 2 m.                  e. more than 16 m.
        c. 4 m.

# 8

# Gravitation

1. According to Kepler's laws, the paths of planets about the                (d)A
   Sun are

   a. parabolas.          d. ellipses.
   b. circles.            e. none of these.
   c. straight lines.

2. What is the force of gravity on a 700-Newton man standing                (c)A
   on the Earth's surface?

   a. 70 N.               c. 709.8 N.
   b. 350 N.              d. none of these.
   c. 700 N.

3. If the mass of the Earth increased with no change in                     (a)A
   radius, your weight would

   a. increase also.      c. stay the same.
   b. decrease.

4. If the radius of the Earth decreased with no change in                   (a)A
   mass, your weight would

   a. increase.           c. decrease.
   b. not change.

5. If the Earth's mass decreased to one-half its original                   (b)A
   mass with no change in radius, then your weight would

   a. decrease to one quarter your original weight.
   b. decrease to one half your original weight.
   c. stay the same.
   d. none of these.

6. The underlying concept in the oscillating theory of the                  (a)A
   universe is

   a. gravitational forces.
   b. the inverse-square law.
   c. energy oscillations.
   d. radioactivity.
   e. radiation.

7. According to Newton's laws a rock and a pebble will fall at               (b)B
   the same acceleration in a gravitational field because

   a. the gravitational force on each are the same.
   b. the gravitational field strength is the same for both.
   c. both of the above.
   d. none of the above.

- 69 -

8. A 400 N woman stands on top of a very tall ladder so she is one Earth radius above the Earth's surface. How much does she weigh?   (b)B

    a. zero.              d. 400 N.
    b. 100 N.          e. none of these.
    c. 200 N.

9. A very massive object 'A' and a less massive object 'B' move toward each other under the influence of mutual gravitation. Which force, if either, is greater;   (c)B

    a. the force on 'A'.     c. both forces are
    b. the force on 'B'.        the same.

10. Two objects move toward each other because of gravitational attraction.  As the objects get closer and closer, the force between them   (a)B

    a. increases.        c. remains constant.
    b. decreases.

11. The force of gravity acting on you will increase if you   (b)B

    a. burrow deep inside the planet.
    b. stand on a planet with a radius that is shrinking.
    c. both of the above.
    d. none of the above.

12. The force of gravity is greatest on you when you are standing   (c)B

    a. just above the Earth's surface.
    b. just below the Earth's surface.
    c. on the Earth's surface.

13. A lunar month is about 28 days.  If the Moon were closer to the Earth than it is now, the lunar month would be   (a)C

    a. less than 28 days.   c. unchanged at 28 days.
    b. more than 28 days.

14. There would be only one ocean tide per 24 hour day if the   (e)C

    a. Earth and Moon were equally massive.
    b. Moon were more massive than the Earth.
    c. Sun's influence on the tides were negligible.
    d. Moon's mass were small - only a few kilograms.
    e. none of these.

15. When the distance between two stars decreases by half, the force between them   (d)B

    a. decreases by one-quarter.
    b. decreases by one-half.
    c. increases to twice as much.
    d. increases to four times as much.
    e. stays the same.

16. The factor most directly responsible for making a black hole invisible is its   (d)B

    a. size.            d. surface escape velocity.
    b. mass.           e. none of these.
    c. color.

17. If the Sun collapsed to a black hole, the resulting gravitational attraction on the Earth would be                                           (c)B

      a. more.           c. the same.
      b. less.

18. If you drop a stone into a hole drilled all the way to the other side of the Earth, the stone will                                           (b)B

      a. stop at the center of the Earth.
      b. speed up until it gets to the center of the Earth.
      c. speed up until it reaches the other side of the Earth.
      d. slow down until it reaches the center.

19. The reason the Moon does not fall into the Earth is that                                           (c)B

      a. the Earth's gravitational field is weak at the Moon.
      b. the gravitational pull of other planets keeps the Moon up.
      c. the Moon has a sufficiently large orbital speed.
      d. the Moon has less mass than the Earth.
      e. none of these.

20. When a star collapses to form a black hole, its mass                                           (c)B

      a. increases.      c. remains the same.
      b. decreases.

21. A hollow spherical planet is inhabited by people who live inside it, where the gravitational field is zero. When a very massive space ship lands on the planet's surface, inhabitants find that the gravitational field inside the planet is                                           (b)C

      a. still zero.
      b. non-zero, directed toward the spaceship.
      c. non-zero, directed away from the spaceship.

22. A supplier wants to make a profit by buying metal by weight at one altitude and selling it at the same price per pound at another altitude. The supplier should                                           (a)C

      a. buy at a high altitude and sell at a low altitude.
      b. buy at a low altitude and sell at a high altitude.
      c. disregard altitude because it makes no difference.

23. Each of us weighs a tiny bit less on the ground floor of a skyscraper than we do on the top floor. The reason for this is                                           (b)C

      a. the gravitational field is shielded inside the building.
      b. the mass of the building attracts you upward slightly.
      c. both of the above.
      d. none of the above.

Chapter 8b: Tides

1. The main reason ocean tides exist is that                          (b)A

    a. the Sun and Moon pull in conjunction at high tides
       and opposition at low tides.
    b. the Moon's pull on oceans closer to the Moon is
       larger than its pull on oceans farther from the
       Moon.
    c. the Moon is closer to the Earth than is the Sun.
    d. the Moon and the Sun pull in opposite directions
       on the oceans.
    e. none of these.


2. Suppose the Moon had twice as much mass as it now does and       (d)B
   still orbits the  Earth at the same distance.  In that
   case,

    a. the ocean tide facing the Moon would be higher, while
       the other tide would be smaller.
    b. the ocean tide facing the Moon would be smaller, while
       the other tide would be higher.
    c. both ocean tides would be smaller.
    d. both ocean tides would be higher.
    e. the ocean tides would be the same as they are now.


3. Suppose the Moon were covered with water and orbits the         (b)B
   earth as it now does.  In that case the Moon would have

    a. one tidal bulge.     d. four tidal bulges.
    b. two tidal bulges.    e. no tidal bulges.
    c. three tidal bulges.


4. During an eclipse of the Sun, when the Moon comes between        (a)C
   the Sun and the Earth, the ocean tide nearest the Moon
   would be

    a. extra high.          c. the same as
    b. extra low.              always.


5. A lunar month is about 28 days.  If the Moon were closer to     (a)C
   the Earth than it is now, the lunar month would be

    a. less than 28 days.   c. 28 days.
    b. more than 28 days.


6. A lunar month is about 28 days.  If the Moon were farther       (a)C
   from the Earth than it is now, the lunar month would be

    a. more than 28 days.   c. 28 days.
    b. less than 28 days.

7. There would be only one ocean tide per day if the                    (e)C

   a. Earth and Moon had equal masses.
   b. Earth had less mass than the Moon.
   c. Sun's influence on the tides was negligible.
   d. Moon had a small mass.
   e. none of these.


8. There would still be two ocean tides per 24 hours if the             (d)C

   a. Earth and Moon had equal masses.
   b. Moon had more mass than the Earth.
   c. Sun's influence on the tides was negligible.
   d. all of the above.
   e. none of these.

# 9

# The Atomic Nature of Matter

1. How many different elements are in a water molecule?                    (b)A

    a. one.             d. four.
    b. two.             e. none of these.
    c. three.

2. There are about as many atoms of air in our lungs at any                (d)A
   moment as there are breaths of air in

    a. a large auditorium.  d. the whole world.
    b. a large city.       e. none of these.
    c. the United States.

3. Assuming all the atoms exhaled by Julius Ceasar in his last            (a)A
   dying breath are still in the atmosphere, then we breathe
   one of those atoms with each

    a. single breath.    d. ten years.
    b. day.            e. month.
    c. ... it depends, some people still breathe a few of
       Ceasar's atoms every day, while others wouldn't
       breathe one for an entire year.

4. Nuclei of atoms that make up a newborn baby were made in                (c)A

    a. the mother's womb.
    b. the food the mother eats before giving birth.
    c. ancient stars.
    d. the Earth.
    e. none of these.

5. Atoms heavier than hydrogen were made by                                (b)A

    a. photosynthesis.    d. radioactivity.
    b. nuclear fusion.    e. none of these.
    c. radiant energy conversion.

6. If we doubled the magnifying power of the most powerful                 (c)A
   optical microscope in the world, we would

    a. be able to see individual atoms.
    b. be able to photograph individual atoms, even though
       we couldn't see them.
    c. still not be able to see an atom.

7. Which of the following atoms is the most massive?                       (d)A

    a. hydrogen.      d. uranium.
    b. iron.          e. ...all have the
    c. lead.            same mass.

8. Which of the following statements are true?                    (d)A

   a. A molecule is the smallest particle that exists.
   b. Chemical elements are made up of 106 distinct
      molecules.
   c. Molecules form atoms which in turn determine chemical
      properties of a substance.
   d. Molecules are the smallest subdivision of matter that
      still retain chemical properties of a substance.
   e. None of these statements is true.


9. Which of the following statements is true?                     (b)A

   a. An atom is the smallest particle known to exist.
   b. There are only 106 different kinds of atoms which
      combine to form all substances.
   c. There are thousands of different kinds of atoms
      which account for a wide variety of substances.
   d. Atoms are so small that there is no way we can photo-
      graph them.
   e. None of these statements is true.


10. What makes an element distinct?                               (a)A

   a. the number of protons.
   b. the number of neutrons.
   c. the number of electrons.
   d. the total mass of all the particles.
   e. none of these.


11. What normally determines whether a substance is a liquid,     (d)A
    solid, gas, or plasma?

   a. The atomic number of the atoms involved.
   b. The atomic mass of the atoms involved.
   c. The density of the substance.
   d. The temperature of the substance.
   e. none of these.


12. Brownian motion has to do with the                            (d)A

   a. size of atoms.
   b. atomic vibrations.
   c. first direct measurement of atomic motion.
   d. random motions of atoms and molecules.
   e. rhythmic movements of Brownians.


13. Which of the following is not a compound?                     (a)A

   a. air.                 d. salt.
   b. ammonia.             e. ... all are compounds.
   c. water.


14. Which of the following is not a mixture?                      (d)A

   a. granite.             d. sugar.
   b. cake.                e. ... all are mixtures.
   c. air.


15. In an electrically neutral atom, the number of protons in the  (c)A
    nucleus is balanced by an equal number of

   a. neutrons.            d. all of the above.
   b. pions.               e. none of the above.
   c. orbital electrons.

16. Which is the smallest particle of the ones listed below?                     (c)A

      a. a molecule.         c. a proton.
      b. an atom.           d. a nucleus.

17. A molecule has                                                               (d)A

      a. mass.            d. all of the above.
      b. structure.     e. none of the above.
      c. energy.

18. The reason a granite block is mostly empty space is because                   (b)A
the atoms in the granite are

      a. in perpetual motion.
      b. mostly empty space themselves.
      c. held together by electrical forces.
      d. not as close together as they could be.
      e. invisible.

19. Solid matter is mostly empty space.  The reason solids don't                 (d)A
fall through one another is because

      a. atoms are constantly vibrating, even at absolute zero.
      b. of nuclear forces.
      c. of gravitational forces.
      d. of electrical forces.
      e. none of the these.

20. Which of these forces determines the chemical properties of                  (d)A
an atom?

      a. friction force.    d. electrical force.
      b. nuclear force.     e. none of these.
      c. gravitational force.

21. The air in this room has                                                     (d)A

      a. mass.            d. all of the above.
      b. weight.        e. none of the above.
      c. energy.

22. What is the molecular mass of a water molecule?                              (d)A

      a. 10 amu.        d. 18 amu.
      b. 12 amu.        e. none of these.
      c. 15 amu.

23. If two protons are removed from an oxygen nucleus, the result                (b)B
is

      a. nitrogen.      d. neon.
      b. carbon.       e. none of these.
      c. helium.

24. If one neutron is added to a helium nucleus, the result is                   (e)B

      a. hydrogen.     d. beryllium.
      b. boron.       e. helium.
      c. lithium.

# 10

## Solids

1. Diamond and graphite have different physical properties          (b)A
   because of their different

    a. elements.        d. both of the above.
    b. electron bondings.  e. none of the above.

2. The fundmental force responsible for all atomic bonding is          (c)A

    a. nuclear.      d. frictional.
    b. chemical.     e. none of these.
    c. electrical.

3. The weakest interatomic bond is          (c)A

    a. covalent.     d. metallic.
    b. ionic.       e. none of these.
    c. Van der Waals.

4. Brass and bronze are examples of          (c)A

    a. elements.    d. compounds.
    b. mixtures.    e. molecules.
    c. alloys.

5. Which of the following is not an alloy?          (b)A

    a. bronze.     d. brass.
    b. copper.    e. ... all are alloys.
    c. steel.

6. When a solid block of material is cut in half, its density is          (b)A

    a. halved.     c. doubled.
    b. unchanged.

7. Which has the greater density, a lake full of water or a cup          (c)A
   full of lake water?

    a. the cup.    c. both have the same
    b. the lake.      density.

8. Compared to the density of a kilogram of feathers, the          (b)A
   density of a kilogram of lead is

    a. less.       c. the same.
    b. more.

9.  If a loaf of bread is compressed, its                                    (d)A

    a. surface tension becomes less.
    b. molecules become harder.
    c. density decreases.
    d. density increases.
    e. none of these.


10. If the mass of an object were to double while its volume      (b)A
    remains the same, its density would

        a. half.                    c. stay the same.
        b. double.


11. If the volume of an object were to double while its mass      (a)A
    stays the same, its density would

        a. half.                    c. stay the same.
        b. double.


12. A wooden block has a mass of 1000 kg and a volume of          (c)AM
    2 meters cubed.  What is the block's density?

        a. 100 kg per meter cubed.
        b. 200 kg per meters cubed.
        c. 500 kg per meters cubed.
        d. 1000 kg per meters cubed.
        e. none of these.


13. A block of iron is placed in a furnace where it is heated and   (a)B
    consequently expands.  In the expanded condition, its density

        a. is less.                 c. is more.
        b. is the same.


14. Which will bounce higher off a hard surface?                   (b)B

        a. a rubber ball.           c. both bounce the same.
        b. a steel ball.


15. When a weight is suspended on a spring, the spring stretches   (c)B
    10 cm.  If the weight is doubled, the spring will stretch

        a. 10 cm.                   d. 40 cm.
        b. 15 cm.                   e. more than 40 cm.
        c. 20 cm.


16. If you hang a one kilogram weight from a spring, the spring    (d)B
    will stretch 10 cm.  If instead you hang the kilgram weight
    from two such springs hanging side by side, so each spring
    supports half the weight, then each spring will stretch

        a. 40 cm.                   d. 5 cm.
        b. 20 cm.                   e. none of these.
        c. 10 cm.


17. A metal block has a density of 5000 kg per meters cubed and    (d)BM
    a volume of 2 cubic meters.  What is the block's mass?

        a. 1000 kg.                 d. 10000 kg.
        b. 2500 kg.                 e. none of these.
        c. 5000 kg.

18. A metal block has a density of 5000 kg per meters cubed and          (b)BM
    a mass of 15,000 kg.  What is its volume?

         a. 0.33 cubic meters.     d. 15 cubic meters.
         b. 3 cubic meters.        e. none of these.
         c. 5 cubic meters.

Chapter 10b: Scaling

1. Which potatoes when peeled produce the most peelings?                    (b)A

     a. 10 kg of large potatoes.
     b. 10 kg of small potatoes.
     c. they both produce the same amount.

2. Compared to a 50 kg person, a 100 kg person at the beach              (c)A
requires

     a. more than twice as much suntan lotion.
     b. the same amount of suntan lotion.
     c. less than twice as much suntan lotion.

3. A kilogram of peaches have more skin area than a kilogram of        (b)A

     a. blueberries.     c. grapes.
     b. grapefruits.     d. ...each has the same skin area.

4. A house husband is making taffy apples.  If he buys 100 kg         (b)A
of small apples rather than 100 kg of large apples, he will
need

     a. less taffy.     c. the same amount of taffy.
     b. more taffy.

5. Doubling the linear size of an object, increases its area by        (b)A

     a. two and its volume by four.
     b. four and its volume by eight.
     c. eight and its volume by sixteen.
     d. none of these.

6. An elephant eats less for its size than smaller animals          (b)A
because

     a. its ears are bigger.
     b. its surface area is small compared to its volume.
     c. its surface area is large compared to its volume.
     d. it is taller than other animals.
     e. it weighs more than smaller animals.

7. Which geometrical shape has the least surface area for a given       (d)A
given volume?

     a. cube.     d. sphere.
     b. pyramid.     e. none of these.
     c. cylinder.

8. Which cooks faster in boiling oil?                                  (b)A

     a. a whole potato.     c. both cook the same.
     b. a sliced potato.

9. A dome-shaped house is more heat efficient than a rectangular     (c)A
   house because a dome has

      a. no corners to radiate.
      b. less inner space.
      c. less area compared to its volume.
      d. all of the above.
      e. none of the above.

10. If an elephant grew to twice its height, the area of             (b)A
    its ears would be about

      a. twice what it was.
      b. four times what it was.
      c. six times what it was.
      d. eight times what it was.
      e. none of these.

11. An elephant that grows to twice its normal height, will          (d)A
    increase its weight by about

      a. two times.      d. eight times.
      b. four times.     e. none of these.
      c. six times.

12. Suppose all sized potatoes are selling at the same price         (c)A
    per kilogram. You will get more peeled potatoes for your
    money if you buy

      a. small potatoes.    c. large potatoes.
      b. medium potatoes.   d. ...makes no difference.

13. If a pencil's length increases by 10 times and its diameter      (b)A
    increases by 10 times, then its surface area increases by

      a. 10.      d. 10,000.
      b. 100.     e. none of these.
      c. 1000.

14. If a pencil's length increases by 10 times and its diameter      (c)A
    increases by 10 times, then its volume increases by

      a. 10.      d. 10,000.
      b. 100.     e. none of these.
      c. 1000.

15. Consider the fictional case of the incredible shrinking          (c)B
    man. If he shrinks proportionately to 1/10 his original
    height, his weight will decrease by

      a. 0.1.     0.0001.
      b. 0.01.    e. none of these.
      c. 0.001.

16. Consider the fictional case of the incredible shrinking          (b)B
    woman. If her height shrinks by 0.1, then her area shrinks
    by

      a. 0.1.     d. 0.0001.
      b. 0.01.    e. none of these.
      c. 0.001.

17. If you make cupcakes and bake them as directed for a cake,       (a)B
    you will find the cupcakes

      a. overbaked.    c. properly baked.
      b. underbaked.

18. In cold weather, your hands will be warmer if you wear                    (b)B

      a. gloves.         c. both will be the same.
      b. mittens.


19. If each dimension of a steel bridge is scaled up ten times,                (b)B
    its strength will increase by

      a. ten and its weight by ten also.
      b. one hundred, but its weight by one thousand.
      c. one thousand, and its weight by one hundred.
      d. none of these.

# 11

# Liquids

Chapter 11a: Liquids

1. Water pressure is greatest against the                                  (b)A

    a. top of a submerged object.
    b. bottom of a submerged object.
    c. sides of a submerged object.
    d. ...is the same against all surfaces.
    e. none of these.


2. A dam is thicker at the bottom than at the top because           (a)A

    a. water pressure is greater with increasing depth.
    b. surface tension exists only on the surface of liquids.
    c. water is denser at deeper levels.
    d. it looks better.
    e. none of these.


3. The pressure in a liquid depends on liquid                          (d)A

    a. density.              d. all of the above.
    b. velocity.             e. none of the above.
    c. depth.


4. The pressure at the bottom of a jug filled with water does          (d)A
   NOT depend on

    a. the acceleration due to gravity.
    b. water density.
    c. the height of the liquid.
    d. surface area of the water.
    e. none of these.


5. Pumice is a volcanic rock that floats. Its density is               (a)A

    a. less than the density of water.
    b. equal to the density of water.
    c. more than the density of water.


6. What is the bouyant force acting on a 10 ton ship floating          (b)A
   in the ocean?

    a. less than 10 tons.    c. more than 10 tons.
    b. 10 tons.              d. depends on density of sea water.


7. What is the weight of water displaced by a 100 ton floating         (b)A
   ship?

    a. less than 100 tons.   d. 100 cubic meters.
    b. 100 tons.             e. depends on the ship's shape.
    c. more than 100 tons.

8. When an object is partly or wholly immersed in a liquid, it is buoyed up                                                                                                      (b)A

    a. by a force equal to its own weight.
    b. by a force equal to the weight of liquid displaced.
    c. and floats because of Archimedes principle.
    d. but nevertheless sinks.
    e. none of these.


9. The buoyant force on an object is least when the object is                                                                                                      (a)A

    a. partly submerged.
    b. submerged near the surface.
    c. submerged near the bottom.
    d. none of these.


10. The reason objects immersed in a fluid experience an upward buoyant force is because fluid pressure on the bottom of the object is greater than fluid pressure on the top of the object.                                                                      (a)A

    a. true.                   b. false.


11. The reason a life jacket helps you float is                                                                                                      (e)A

    a. the jacket makes you weigh less.
    b. the jacket has the same density as an average human.
    c. the jacket repels water.
    d. if you sink, the jacket sinks.
    e. you and the jacket together have density less than
       your density alone.


12. A jar contains 200 N of water.  The area of the inside bottom of the jar is 2 square meters.  What pressure does the water exert on the bottom of the jar?                                                            (a)AM

    a. 100 N per square meter.
    b. 200 N per square meter.
    c. 400 N per square meter.
    d. 800 N per square meter.
    e. none of these.


13. Lobsters live on the bottom of the ocean.  The density of a lobster is                                                                                                      (a)B

    a. greater than the density of sea water.
    b. equal to the density of sea water.
    c. less than the density of sea water.


14. The density of a submerged submarine is about the same as the density of                                                                                                      (d)B

    a. a crab.                 d. water.
    b. iron.                   e. none of these.
    c. a floating submarine.


15. A rock weighs 5 N in air and 3 N in water.  What is the buoyant force on the rock?                                                                                                      (d)B

    a. 8 N.                    d. 2 N.
    b. 5 N.                    e. none of these.
    c. 3 N.

16. An egg is placed at the bottom of a bowl filled with water.    (b)B
    Salt is slowly added to the water until the egg rises and
    floats.  From this experiment, one concludes

    a. calcium in the egg shell is repelled by sodium cloride.
    b. the density of salt water exceeds the density of egg.
    c. buoyant force does not always act upward.
    d. salt sinks to the bottom.
    e. none of these.

17. Ice cubes submerged at the bottom of a liquid mixture indicate    (e)B
    that the mixture

    a. fails to produce a buoyant force on the ice.
    b. has dissolved air in a liquid state.
    c. is composed of open-structured crystals.
    d. is not displaced by the submerged ice.
    e. is less dense than ice.

18. A block of wood weighs 5 N in air.  The buoyant force that    (b)B
    mercury will exert on the wood as it floats in mercury is

    a. less than 5 N.        c. more than 5 N.
    b. 5 N.

19. The volume of water displaced by a floating 20 ton boat is    (b)B

    a. 20 cubic meters.
    b. the volume of 20 tons of water.
    c. the volume of the boat.
    d. ...depends on the shape of the ship's hull.
    e. none of these.

20. Compared to an empty ship, the same ship loaded with    (b)B
    styrofoam will float

    a. higher in the water.
    b. lower in the water.
    c. at the same level in the water.

21. Two equal sized buckets are filled to the top with water.    (b)B
    One of the buckets has a piece of wood floating in it,
    making its total weight

    a. less than the weight of the other bucket.
    b. equal to the weight of the other bucket.
    c. more than the weight of the other bucket.

22. A block of balsa wood foats on water while a same size block    (a)B
    of lead lies submerged in the water.  The buoyant force is
    greatest on the

    a. lead.              c. ...is the same for both.
    b. wood.

23. A liter-sized block of ordinary wood is put in water.  The    (a)B
    amount of water displaced is

    a. less than 1 liter.    d. depends on the water density.
    b. 1 liter.              e. none of these.
    c. more than 1 liter.

24. A block of lead and a same-sized block of aluminum are    (c)B
    submerged in water.  Buoyant force is greatest on the

    a. lead.          '     c. ...same on each.
    b. aluminum.

25. A wooden cube exerts a pressure of 200 N per square meter          (b)BM
    on a table top.  Each side of the cube has an area of 0.5
    meters squared.  How much force is the cube exerting on the
    table top?

        a. 40 N.                d. 1000 N.
        b. 100 N.               e. none of these.
        c. 200 N.

26. A wooden cube weighs 20 N and exerts a pressure of 100 N           (b)BM
    per meter squared on a table top.  What is the area of a
    side of the cube?

        a. 0.1 meters squared.  d. 20 meters squared.
        b. 0.2 meters squared.  e. 100 meters squared.
        c. 5.0 meters squared.

27. A kilogram of lead and a kilogram of aluminum are submerged        (b)C
    in water.  Buoyant force is greatest on the

        a. lead.                c. ...same on each.
        b. aluminum.

28. A boat loaded with a barrel of water floats in a swimming          (c)C
    pool.  When the water in the barrel is poured overboard, the
    swimming pool level will

        a. rise.                c. remain unchanged.
        b. fall.

29. A boat loaded with scrap iron floats in a swimming pool.           (b)C
    When the iron is thrown overboard, the pool level will

        a. rise.                c. remain unchanged.
        b. fall.

30. A boat loaded with wood floats in a swimming pool.  When the       (c)C
    wood is thrown overboard, the pool level will

        a. rise.                c. remain unchanged.
        b. fall.

31. If a battleship sinks in a canal lock, the water level in          (b)C
    the lock will

        a. rise.                c. remain unchanged.
        b. fall.

32. When a boat sails from fresh water to salt water, the boat         (b)C
    will float

        a. lower in the water.  c. at the same water
        b. higher in the water.    level.

33. If the part of an iceberg that extends above the water were        (b)C
    suddenly removed, the

        a. iceberg would sink.
        b. buoyant force on the iceberg would decrease.
        c. density of the iceberg would change.
        d. pressure on the bottom of the iceberg would increase.
        e. none of these.

34. If a weighted air filled balloon sinks in a lake, it will          (d)C

      a. probably sink to the bottom and probably rise later.
      b. sink until it reaches equilibrium and then remain at
         constant depth.
      c. be buoyed up with constant force while sinking.
      d. always sink to the bottom.
      e. none of these.


35. If a weighted air filled balloon sinks in water, it will          (d)C

      a. sink to an equilibrium level and then stop.
      b. burst if water pressure is great enough.
      c. become less dense as it sinks.
      d. be acted on by a continually decreasing buoyant force.
      e. none of these.


36. As a weighted air filled balloon sinks in water,          (d)C

      a. its volume increases.
      b. it has a greater chance of bursting.
      c. it will reach an equilibrium level.
      d. its buoyancy decreases.
      e. none of these.


37. When an ice cube in a glass of water melts, the water level          (c)C

      a. rises.              c. remains the same.
      b. falls.


38. A floating ice cube contains a small piece of iron.  After          (b)C
    melting the water level will

      a. rise.               c. remain unchanged.
      b. fall.


39. An ice cube floating in a glass of water contains many air          (c)C
    bubbles.  When the ice melts, the water level will

      a. rise.               c. remain unchanged.
      b. fall.


40. Three icebergs are each floating in bathtubs filled to the          (a)C
    brim with water.  Iceberg A has large air bubbles in it.  Ice-
    berg B has unfrozen water in it.  Iceberg C has an iron rail-
    road spike in it.  When the ice melts, what happens?

      a. the water level in C will decrease while the other
         two water levels will remain the same.
      b. only the water in C will spill over.
      c. the water level in A will stay the same, while the
         other tubs will spill over.
      d. all the tubs spill over.
      e. all stay exactly the same.


41. A block of wood floats partly out of the water.  A piece of          (a)C
    iron is placed on top of the wood so it floats with the water
    level even with the top of the wood.  If the iron is suspended
    beneath the wood, then the top face of the wood will be

      a. above the water.     c. even with the water line.
      b. beneath the water.

42. There is a legend of a Dutch boy who bravely held back the        (c)C
   Atlantic Ocean by plugging a leak near the top of a dike with
   his finger until help arrived.  Which of the following is
   most likely?

    a. This is impossible because of the large size of the
          Atlantic Ocean.
    b. The force on the his finger would have been huge, but
          the pressure small enough for this to happen.
    c. The force on his finger would have been less than 1 N.
    d. Both the force and pressure on his finger would have
          been great, but the boy was very strong.
    e. none of these.

1. A hydraulic arrangement consists of a water filled U-tube                    (b)A
   that is wider on one end than on the other.  Pistons are
   fitted in both ends.  To multiply an input force, the input
   end should be the one having the

   a. larger diameter piston.
   b. smaller diameter piston.
   c. ...it doesn't matter what relative size the
      piston is.

2. A hydraulic arrangement consists of a water filled U-tube                    (b)A
   that is wider on one end than on the other.  Pistons are
   fitted in both ends.  The ratio of output force to input
   force will be equal to the ratio of the output and input
   piston

   a. diameters.             d. all of the above.
   b. areas.                 e. none of the above.
   c. radii.

3. A hydraulic press multiplies a force by 100.  This multipli-                 (b)A
   cation is done at the expense of

   a. energy, which decreases by a factor of 100.
   b. the distance through which the force acts.
   c. the time through which the force acts, which is
      extended by a factor of 100.
   d. the mechanism providing the force.
   e. none of these.

4. In a hydraulic press operation, it is impossible for the                     (c)A

   a. output displacement to exceed the input displacement.
   b. force ouput to exceed the force input.
   c. energy output to exceed the energy input.
   d. output piston's speed to exceed the input piston's
      speed.
   e. none of these.

5. Surface tension is a direct result of                                        (e)A

   a. adhesive forces between molecules in a liquid or solid.
   b. viscosity.
   c. Archimedes principle.
   d. Bernoulli's principle.
   e. cohesive forces between molecules in a liquid.

6. A consequence of the surface tension of water is                             (d)A

   a. capillary action.
   b. wet sand being firmer than dry sand.
   c. hot oily soup tasting different than cold oily soup.
   d. all of the above.
   e. none of the above.

7. When you put a stick in water and remove it, the stick is wet.    (a)A
When you put a stick in mercury and remove it, the stick is
dry.  The reason for this is adhesive forces are greater
between the stick and

   a. water.              c. not enough information
   b. mercury.               given.


8. A very light-weight loop of wire is suspended from a fine     (a)A
spring, lowered into water, and then raised to the surface.
Any further attempt to raise it causes the spring to

   a. stretch.            c. stay the same.
   b. contract.


9. Surface tension of liquids                                    (b)B

   a. increases when wetting agents are added.
   b. decreases as the liquid temperature increases.
   c. is about the same for all liquids.
   d. is actually a thin membrane on the liquid surface that
      consists of liquid and air molecules.
   e. is the reason a steel ship will float.

# 12

# Gases and Plasmas

Chapter 12a: Gases

1. A balloon is buoyed up with a force equal to the       (a)A

    a. weight of air it displaces.
    b. density of surrounding air.
    c. atmospheric pressure.
    d. weight of the balloon and contents.
    e. all of these.

2. A bubble of air released from the bottom of a lake    (c)A

    a. rises to the top at constant volume.
    b. becomes smaller as it rises.
    c. becomes larger as it rises.
    d. alternately expands and contracts as it rises.
    e. none of these.

3. A one ton dirigible hovers in the air.  The buoyant force  (b)A
acting on it is

    a. zero.          c. less than one ton.
    b. one ton.      d. more than one ton.

4. A 5 liter metal can will float in air if it is    (e)A

    a. evacuated of air.
    b. filled with many Newtons of air pressure.
    c. thrown high enough into the air.
    d. filled with enough helium.
    e. ...nonsense! A metal can, whatever its contents
       can never float in air.

5. In drinking soda water through a straw, one uses   (c)A

    a. capillary action.   d. Bernoulli's principle.
    b. surface tension.    e. none of these.
    c. atmospheric pressure.

6. If you are standing on the moon and release a balloon filled (b)A
with helium gas, the balloon will

    a. rise.         c. neither rise nor
    b. fall.           fall.

7. The air in this room has             (d)A

    a. mass.         d. all of the above.
    b. weight.      e. none of the above.
    c. energy.

8. The faster a fluid moves, the          (b)A

    a. greater its internal pressure.
    b. lesser its internal pressure.

9. When water is turned on in a shower, the shower curtain moves     (d)A
   towards the water.  This has to do with

        a. capillary action.    d. pressure in a moving fluid.
        b. surface tension.     e. none of these.
        c. heat capacity.


10. On a windy day, atmospheric pressure                             (b)A

        a. increases.           c. remains unchanged.
        b. decreases.


11. If a TV picture tube is broken, it will                          (b)A

        a. explode.             c. smoke and flame.
        b. implode.             d. none of these.


12. About how high can water be theoretically lifted by a            (c)A
    vacuum pump at sea level?

        a. less than 10.3 m.    c. 10.3 m.
        b. more than 10.3 m.


13. One barometer tube has twice the cross-sectional area of         (a)A
    another.  Mercury in the smaller tube will rise

        a. the same height as in the larger tube.
        b. twice as high as mercury in the larger tube.
        c. four times as high as mercury in the larger tube.
        d. more than four times as high as in the larger tube.
        e. none of these.


14. Gas pressure inside an inflated stretched balloon is             (c)B

        a. less than air pressure outside the balloon.
        b. equal to air pressure outside the balloon.
        c. greater than air pressure outside the balloon.


15. As a high altitude balloon sinks lower and lower into the        (a)B
    atmosphere, its

        a. volume decreases.    d. mass decreases.
        b. density decreases.   e. none of these.
        c. weight decreases.


16. As a balloon rises higher and higher into the atmosphere, its    (e)B

        a. volume decreases.    d. mass decreases.
        b. density increases.   e. none of these.
        c. weight increases.


17. A helium filled balloon released in the atmosphere will rise     (c)B
    until

        a. the pressure inside the balloon equals atmospheric
           pressure.
        b. atmospheric pressure on the bottom of the balloon
           equals atmospheric pressure on the top of the balloon.
        c. the balloon's density equals the atmospheric density.
        d. all of these.
        e. none of these.

18. A helium filled balloon released in the atmosphere will continue     (e)B
    to rise until

    a. the pressure inside the ballon equals atmospheric
       pressure.
    b. atmospheric pressure on the bottom of the balloon
       equals atmospheric pressure on the top of the balloon.
    c. the balloon can no longer expand.
    d. all of these.
    e. none of these.

19. Two tubes of equal cross sectional area are filled with water     (c)B
    and mercury.  The heights of the liquids are 10.3 m and 0.76 m.
    Both liquids have equal

    a. volumes.                d. number of molecules.
    b. densities.              e. none of these.
    c. weights.

20. Alcohol is less dense than water.  If alcohol is used to     (b)B
    make a barometer on a day when atmospheric pressure is normal,
    the level of the alcohol would be

    a. less than 10.3 m.    c. 10.3 m.
    b. more than 10.3 m.

21. Compared to the buoyant force of the atmosphere on a one     (c)B
    liter helium filled balloon, the buoyant force of the atmo-
    sphere on a near-by one liter solid iron block is

    a. considerably less.    c. the same.
    b. considerably more.

22. Compared to the buoyant force of the atmosphere on a one     (a)B
    kilogram helium filled balloon, the buoyant force of the atmo-
    sphere on a near-by one kilogram solid iron block is

    a. considerably less.    c. the same.
    b. considerably more.

23. As a woman holding her breath swims deeper and deeper beneath     (a)B
    the water's surface, her density

    a. increases.            c. remains the same.
    b. decreases.

24. When a container of gas is squeezed to half its volume, its     (b)B
    density

    a. halves.               c. quadruples.
    b. doubles.              d. remains the same.

25. A swimmer cannot snorkel more than a meter deep.  The reason     (d)B
    for this is, at that depth the lungs are so compressed that

    a. air in the lungs cannot easily be expelled.
    b. air tends to liquify in the snorkel tube.
    c. they will be buoyed up leaving the swimmer breathless.
    d. low pressure air at the surface will not enter the
       higher pressure area in the lungs.
    e. all of these.

26. When a gas container is squeezed to half its volume and the     (b)B
    temperature remains the same, the gas pressure

    a. halves.               c. quadruples.
    b. doubles.              d. remains the same.

27. An umbrella tends to move upwards on a windy day because         (d)B

    a. air gets trapped under the umbrella and pushes it up.
    b. buoyancy increase with increasing wind speed.
    c. winds pushes it up.
    d. a low pressure area is created on top of the umbrella.
    e. all of these.

28. Wind blowing over the top of a hill         (b)B

    a. increases atmospheric pressure there.
    b. decreases atmospheric pressure there.
    c. does not affect atmospheric pressure there.

29. The Bernoulli effect causes passing ships to be drawn together     (c)B
when the ships are close and moving in the

    a. same direction.     c. either the same or
    b. opposite direction.      opposite direction.

30. It would be easier to pull evacuated Magdeburg hemispheres     (d)B
apart when they are

    a. held upside down.
    b. at sea level.
    c. 20 km beneath the ocean surface.
    d. 20 km above the ocean surface.
    e. none of these.

31. In a vacuum, an object has no         (a)B

    a. buoyant force.     d. weight.
    b. mass.     e. temperature.
    c. heat.

32. A car with closed windows makes a left hand turn.  A helium     (b)C
filled balloon in the car will move to the

    a. right.     c. front.
    b. left.     d. back.

33. An empty jar is pushed open-side downward into water so     (b)C
that air trapped in it cannot get out.  As it is pushed
deeper, the buoyant force on the jar

    a. increases.     c. remains the same.
    b. decreases.

Chapter 12b: Plasmas

1. A gas and a plasma are similiar in that they are both          (d)A

    a. crystalline.      c. ionic.
    b. covalent.       d. fluids.


2. When a gas is heated and becomes a plasma, its electric          (a)A
charge is usually

    a. balanced.      d. non-existent
    b. negative.      e. none of these.
    c. positive.


3. Which of the following is an example of matter in a plasma          (b)A
state?

    a. dry ice.        d. liquid hydrogen.
    b. a toroh flame.    e. none of these.
    c. molten lava.


4. Compared to all liquids, solids, and gases in the universe,          (a)A
plasmas are the most

    a. abundant.      c. ...we don't have enough
    b. rare.          information at this time.


5. The main difference between gases and plasmas has to do with          (c)A

    a. the kinds of elements involved.
    b. interatomic spacing.
    c. electrical conduction.
    d. fluid pressure.
    e. the proportion of matter to antimatter in the universe.


6. When a common fluorescent lamp is on, the mercury vapor          (c)A
inside is actually in a

    a. gaseous state.    d. solid state.
    b. liquid state.     e. none of these.
    c. plasma state.


7. Most of the matter in the universe is in the          (d)A

    a. solid state.     d. plasma state.
    b. liquid state.    e. none of these.
    c. gaseous state.

# 13

# Temperature, Heat, and Expansion

1. When an iron ring is heated, the hole becomes                               (b)A

   a. smaller.          c. neither smaller
   b. larger.              nor larger.

2. When a bimetallic bar made of copper and iron strips is          (c)A
   heated, the bar bends toward the iron strip.  The reason
   for this is

   a. iron gets hotter before copper.
   b. copper gets hotter before iron.
   c. copper expands more than iron.
   d. iron expands more than copper.
   e. none of these.

3. During a very cold winter, water pipes sometimes burst.  The     (b)A
   reason for this is

   a. the ground contracts when colder, pulling pipes apart.
   b. water expands when freezing.
   c. water contracts when freezing.
   d. the thawing process releases pressure on the pipes.
   e. none of these.

4. Which of the following has the lowest specific heat?             (b)A

   a. water.          d. wood.
   b. iron.           e. cork.
   c. glass.

5. Ice has a lower density than water because ice                   (d)A

   a. sinks.
   b. molecules are more compact in the solid state.
   c. molecules vibrate at lower rates than water molecules.
   d. is made of open-structured, hexagonal cyrstals.
   e. the density of ice decreases with decreasing
      temperature.

6. Compared to a giant iceberg, a hot cup of coffee has more        (b)A

   a. thermal energy and higher temperature.
   b. temperature, but less thermal energy.
   c. specific heat and more thermal energy.
   d. none of these.

7. As a piece of metal with a hole in it cools, the diameter        (b)A
   of the hole

   a. increases.          c. remains the same.
   b. decreases.

8. If glass expanded more than mercury, then the column of       (b)B
   mercury in a mercury thermometer would rise when the
   temperature

      a. increases.       c. neither increases
      b. decreases.         nor decreases.

9. A volume of air has a temperature of 0 degrees Celsius.  An     (d)B
   equal volume of air that is twice as hot has a temperature
   of

      a. 0 degrees C.     d. 273 degrees C.
      b. 2 degrees C.     e. none of these.
      c. 100 degrees C.

10. Compared to the water temperature at the top of a             (b)B
    waterfall, the water temperature at the bottom of the
    waterfall will be

      a. less than at the top.
      b. greater than at the top.
      c. the same as at the top.

11. Some molecules are able to absorb large amounts of energy     (b)B
    in the form of internal vibrations and rotations.
    Materials composed of such molecules would have

      a. low specific heats.  b. high specific heats.

12. The fact that desert sand is very hot in the day and very     (a)B
    cold at night is evidence that sand has a

      a. low specific heat.   b. high specific heat.

13. Which of the following expands most when the temperature is   (d)B
    increased?

      a. iron.        d. helium.
      b. wood.       e. all expand the same.
      c. ice water.

14. Which of the following expands most when the temperature      (d)B
    is lowered?

      a. iron.        d. water at 4 degrees C.
      b. wood.       e. none expands when the
      c. helium.         temperature is lowered.

15. Which of the following contracts most when the temperature    (d)B
    is decreased?

      a. iron.        d. helium.
      b. wood.       e. all contract the same.
      c. water.

16. Which of the following contracts most when the temperature    (c)B
    is increased?

      a. iron.        d. helium.
      b. wood.       e. none of these contract
      c. ice water.       when heated.

17. Consider a sample of water at 6 degrees C.  If the            (a)B
    temperature is increased, the volume of water

      a. expands.      c. remains the same.
      b. contracts.

18. When water at 4 degrees C. is heated it expands.  When          (b)B
    water  at 4 degrees C. is cooled, it

        a. contracts.              c. neither contracts
        b. expands.                   nor expands.


19. A piece of iron and a cup of water both have the same          (b)C
    temperature.  If they are heated so the thermal energy of
    each doubles,

        a. the water will have the higher temperature.
        b. the iron will have the higher temperature.
        c. both will have the same temperature.
        d. not enough information given to say.


20. If the specific heat of water were lower than it is, in          (a)C
    winter, ponds would be

        a. more likely to freeze.
        b. less likely to freeze.
        c. neither more nor less likely to freeze.


21. Consider a piece of metal that is at 10 degrees C.  If it          (c)CM
    is heated until it has twice the thermal energy, its
    temperature will be

        a. 20 degrees C.          d. 566 degrees C.
        b. 273 degrees C.         e. none of these.
        c. 293 degrees C.


22. A container of air is at atmospheric pressure and  27          (c)CM
    degrees C.  To double the pressure in the container, it
    should be heated to

        a. 54 degrees C.          d. 600 degrees C.
        b. 300 degrees C.         e. none of these.
        c. 327 degrees C.

# 14

# Heat Transfer

Chapter 14: Heat Transfer

1. Energy transfer by convection is primarily restricted to    (d)A

    a. solids.         d. fluids.
    b. liquids.       e. none of these.
    c. gases.

2. A good heat conductor is a    (a)A

    a. poor insulator.    c. neither a poor nor
    b. good insulator.      a good insulator.

3. Warm air rises because faster-moving molecules tend to move    (c)A
to regions of less

    a. density.       c. both of the above.
    b. pressure.      d. none of the above.

4. Newton's law of cooling applies to objects that are    (c)A

    a. cooling.       c. both of the above.
    b. heating.       d. none of the above.

5. The silver coating on the glass surfaces of a Thermos bottle    (c)B
reduces energy that is transferred by

    a. conduction.    d. friction.
    b. convection.    e. none of these.
    c. radiation.

6. If a volume of air is warmed, it expands. If a volume of air    (b)D
expands, it

    a. warms.        c. neither warms nor
    b. cools.         cools.

7. When a volume of air is compressed, its temperature    (a)B

    a. increases.     c. neither increases
    b. decreases.     nor decreases.

8. If a poor absorber of radiation were a good emitter, its    (a)B
temperature would be

    a. less than its surroundings.
    b. more than its surroundings.
    c. the same as its surroundings.

9. A good absorber of radiation is                                          (a)B

    a. a good emitter of radiation.
    b. a poor emitter of radiation.
    c. a good reflector.
    d. none of these.


10. A good reflector of radiation is                                        (c)B

    a. a good absorber of radiation.
    b. a good emitter of radiation.
    c. a poor absorber of radiation.
    d. none of these.


11. If you were caught in freezing weather with only a candle for           (a)B
    heat, you would be warmer in

    a. an igloo.            c. a wooden house.
    b. a tent.              d. a car.


12. If molecules in a sample gas moved so they completely missed            (c)B
    each other, the gas's temperature would

    a. increase.           c. stay the same.
    b. decrease.


13. Hydrogen and oxygen molecules in a sample gas have the same             (d)C
    temperature.  This means the hydrogen molecules, on the
    average, have the same

    a. speed and the same kinetic energy.
    b. speed, but more kinetic energy.
    c. speed, but less kinetic energy.
    d. kinetic energy, but more speed.
    e. kinetic energy, but less speed.


14. Suppose you are served coffee at a restaurant before you are            (a)C
    ready to drink it.  In order for it to be the hottest when
    you are ready for it, you should add cream

    a. right away.         c. When you are ready
    b. at any time.           to drink the coffee.


15. Suppose you want to save energy and you're going to leave your          (c)C
    warm house for a half hour on a cold day.  You should turn
    the thermostat

    a. down a little.      c. off.
    b. up a little.        d. to room temperature.

# 15
# Change of State

Chapter 15: Change of State

1. Evaporation is a cooling process and condensation is a                          (a)A

    a. warming process.
    b. cooling process also.
    c. neither a warming nor cooling process.

2. Evaporation is a cooling process because                                        (c)A

    a. heat is radiated during the process.
    b. of conduction and convection.
    c. the more energetic molecules are able to break away
       from the liquid.
    d. the temperature of the remaining liquid decreases.
    e. none of these.

3. Steam burns are more damaging than burns caused by boiling              (c)A
   water because

    a. steam is a vapor of water molecules.
    b. steam occupies more space than water.
    c. steam has more energy per kg than boiling water.
    d. steam has a higher temperature than boiling water.
    e. none of these.

4. When a gas changes to a liquid state, the gas                                   (a)A

    a. releases energy.        c. neither releases nor
    b. absorbs energy.            absorbs energy.

5. When a solid changes to a liquid state, the solid                               (b)A

    a. releases energy.        c. neither releases nor
    b. absorbs energy.            absorbs energy.

6. Which would burn the most?                                                      (b)A

    a. 100 g of water at 100 degrees C.
    b. 100 g of steam at 100 degrees C.
    c. both would be equally damaging.

7. As atmospheric pressure increases, the boiling temperature of           (b)A
   a liquid

    a. decreases.              c. is 100 degree C.
    b. increases.              d. remains unchanged.

8. When liquids change to a solid state, they                                      (b)A

    a. absorb energy.          c. neither absorb nor
    b. release energy.            release energy.

9. An air sample contains 25 % the amount of water vapor it is    (d)A
   capable of holding.  The air's relative humidy is

        a. 125 %.              d. 25 %.
        b. 75 %.               e. none of these.
        c. 50 %.


10. In the mountains, water boils at                               (b)A

        a. a higher temperature than at sea level.
        b. a lower temperature than at sea level.
        c. the same temperature as at sea level.


11. A refrigerator                                                 (c)B

        a. produces cold.
        b. causes heat to disappear.
        c. removes heat from inside the refrigerator.
        d. changes heat into cold.
        e. none of these.


12. The cooling effect inside a refrgerator is produced by         (c)B

        a. an electric motor that converts electrical energy into
           thermal energy.
        b. compressing the refrigeration gas into a liquid.
        c. vaporizing the refrigeration liquid.
        d. proper insulation.
        e. none of these.


13. When boiling water in the mountains, the time needed to reach  (a)B
    the boiling point is

        a. less than at sea level.
        b. more than at sea level.
        c. the same as at sea level.


14. When snow forms in clouds, the surrounding air                 (a)B

        a. warms.              c. neither warms
        b. cools                  nor cools.


15. When snow forms, the air becomes warmer because                (c)B

        a. snow gives off energy as it falls.
        b. air absorbs energy from the sun.
        c. the change from a vapor state to a solid state is a
           process than gives off energy.
        d. snow gives off energy as it melts.
        e. none of these, because the air doesn't warm when
           snow forms.


16. Food cooked in boiling water in the mountains, cooks slower    (c)B
    than when cooked at sea level.  If the temperature under the
    pot is increased, the food will cook

        a. faster.            c. at the same rate
        b. slower.               as before.


17. If you want to cook boiled eggs while in the mountains, you    (c)B
    should

        a. use a hotter flame.
        b. boil the eggs for a shorter time because the boiling
           temperature is less in the mountains.
        c. boil the eggs for a longer time.
        d. not even consider boiling eggs because the water tem-
           perature won't even get hot enough to cook the eggs.
        e. none of these.

18. When water vapor condenses on the inside of a window, the room    (a)B
    becomes slightly

        a. warmer.              c. ...neither warmer
        b. cooler.                 nor cooler.


19. Melting snow                                                       (b)B

        a. warms the surrounding air.
        b. cools the surrounding air.
        c. neither warms nor cools the surrounding air.


20. When heat is added to boiling water, its temperature              (c)B

        a. increases.          c. stays the same.
        b. decreases.


21. On a humid day, water condenses on the outside of a glass         (d)B
    of ice water.  This phenomenon occurs mainly because of

        a. the porousity of glass.
        b. capillary action.
        c. adhesion of water molecules to glass.
        d. the saturation of cooled air.
        e. evaporation.


22. An inventor discovers a harmless and tasteless salt, which,       (b)B
    when added to water changes its boiling point.  The market
    value for this salt will be best if the salt

        a. lowers the boiling point of water.
        b. raises the boiling point of water.
        c. either raises or lowers the boiling point, as the food
           would be cooked either way.


23. The phenomenon of regelation depends on the                       (c)B

        a. freezing point of water.
        b. melting point of water.
        c. open structured nature of ice crystals.
        d. high specific heat of water.


24. Morning dew on the grass is a result of                           (c)B

        a. evaporation of water.
        b. the open-structured form of water crystals.
        c. slow moving molecules coalescing on the grass.
        d. air pressure on water vapor.
        e. none of these.


25. Ice is put in a cooler in order to cool the contents.  In         (d)B
    order to speed up the cooling process, you should

        a. wrap the ice in newspaper.
        b. drain ice water from the cooler periodically.
        c. keep the ice and food well separated.
        d. crush the ice.
        e. light a small fire under one end, causing air currents
           to form inside the cooler.


26. If you were to walk barefoot on hot coals without harming         (a)B
    your feet, it would be best if your feet are

        a. wet.                 c. either wet or dry -
        b. dry.                    it makes no difference.

# 16

# Thermodynamics

1. It is possible to wholly convert a given amount of heat     (b)A
energy into mechanical energy.

    a. true.         b. false.

2. It is possible to totally convert a given amount of     (a)A
mechanical energy into heat.

    a. true.         b. false.

3. The greater the difference in temperature between the input     (a)A
reservoir and the output reservoir for a heat engine, the

    a. greater the efficiency.
    b. less the efficiency.
    c. ...neither, efficiency doesn't depend on temperature
       difference.

4. The first law of thermodynamics is a restatement of     (d)A

    a. the principle of entropy.
    b. the law of heat addition.
    c. the Carnot cycle.
    d. conservation of energy.
    e. none of these.

5. Systems that are left alone, tend to move towards a state of     (b)A

    a. less entropy.     c. no entropy.
    b. more entropy.

6. Compared to the upper air temperature, in a temperature     (a)A
inversion, the lower air temperature is

    a. cooler.        c. the same.
    b. warmer.       d. upside down.

7. Entropy measures     (e)A

    a. temperature at constant pressure.
    b. temperature at constant volume.
    c. temperature as pressure increases.
    d. temperature as volume increases.
    e. messyness.

8. Entropy is closely related to the     (b)A

    a. 1st law of thermodynamics.
    b. 2nd law of thermodynamics.
    c. both of the above.
    d. none of the above.

9. Because of the absence of clouds, weather, and night time,    (d)A
   more energy falls on a solar panel when in orbit above the
   Earth's atmosphere than on the Earth's surface. About how
   much more energy is available in orbit than on the Earth's
   surface?

   a. the same.              d. eight times as much.
   b. twice as much.         e. none of these.
   c. four times as much.

10. Heating the Earth's atmosphere is caused by energy           (c)A

    a. production.           c. both of the above.
    b. consumption.          d. none of the above.

11. Industrialization on Earth eventually heats our              (b)A
    surroundings to temperatures that are higher than usual.
    Industrialization of space, in habitats using exclusively
    solar power, eventually will pollute the solar system.

    a. true.                 b. false.

12. One hundred Joules of heat is added to a system that         (b)AM
    performs 60 Joules of work. The internal energy change of
    the system is

    a. 0 J.                  d. 100 J.
    b. 40 J.                 e. none of these.
    c. 60 J.

13. Suppose you rapidly stir some raw eggs with an eggbeater.    (a)B
    The temperature of the eggs will

    a. increase.             c. remain unchanged.
    b. decrease.

14. Suppose the temperature of the input reservoir in a heat     (a)B
    engine doesn't change. As the sink temperature is
    lowered, the engine's efficiency

    a. increases.            c. stays the same.
    b. decreases.

15. A heat engine would have 100 percent efficiency if its       (d)B
    input reservoir were

    a. 100 times hotter than the exhaust sink.
    b. 1000 times hotter than the exhaust sink.
    c. 100 times cooler than the exhaust sink.
    d. any finite temperature, but the exhaust sink must
       be at absolute zero.
    e. none of these.

16. The lowest temperature possible in nature is                 (b)B

    a. 0 degrees C.          c. 4 K.
    b. -273 degrees C.

17. When mechanical work is done on a system, there can be an    (d)B
    increase in its

    a. thermal energy.       d. all of the above.
    b. internal energy.      e. none of the above.
    c. temperature.

18. When work is done by a system and no heat added to it, the          (b)B
    temperature of the system

         a. increases.              c. remains unchanged.
         b. decreases.


19. An adiabatic process is characterized by the absence of            (c)B

         a. entropy.               d. temperature changes.
         b. pressure changes.      e. none of these.
         c. heat exchanges.


20. When a volume of air is compressed and no heat enters or           (a)B
    leaves, the air temperature will

         a. increase.              c. remain unchanged.
         b. decrease.


21. When a volume of air expands against the environment and no        (b)B
    heat enters or leaves, the air temperature will

         a. increase.              c. remain unchanged.
         b. decrease.


22. Two identical blocks of iron, one at 10 degrees C and the          (b)B
    other at 20 degrees C, are put in contact.  Suppose the
    cooler block cools to 5 degrees C and the warmer block
    warms to 25 degrees C.  This would violate the

         a. 1st law of thermodynamics.
         b. 2nd law of thermodynamics.
         c. both of the above.
         d. none of the above.


23. Suppose you put a closed, sealed can of air on a hot stove         (d)B
    burner.  The contained air will undergo an increase in

         a. internal energy.       d. all of the above.
         b. temperature.           e. none of the above.
         c. pressure.


24. As a blob of air at a high elevation sinks to a lower             (a)B
    elevation, its temperature usually

         a. increases.             c. remains the same.
         b. decreases.


25. If a blob of air is swept upward, its temperature usually         (b)B

         a. increases.             c. remains the same.
         b. decreases.


26. Adiabatic processes occur in the                                  (d)B

         a. atmosphere.            d. all of the above.
         b. oceans.                e. none of the above.
         c. Earth's mantel.


27. The amount of energy ultimately converted to heat by a            (e)B
    light bulb is about

         a. 15 %.                  d. 60 %.
         b. 30 %.                  e. 100 %.
         c. 45 %.

28. If you run a refrigerator in a closed room with the                     (a)B
    refrigerator door open, the room temperature will

        a. increase.           c. remain unchanged.
        b. decrease.

29. As a system becomes more disordered, entropy                            (a)B

        a. increases.          c. remains the same.
        b. decreases.

30. On a chilly 10 degrees C day, your friend who likes cold                (d)BM
    weather, wishes it were twice as cold.  Taken literally,
    this temperature would be

        a. 10 degrees C.       d. -131.5 degrees C.
        b. 5 degrees C.        e. -141.5 degrees C.
        c. 0 degrees C.

31. The ideal efficiency for a heat engine operating between                (d)CM
    temperatures of 2700 K and 300 K is

        a. 10 %.               d. 89 %.
        b. 24 %.               e. none of these.
        c. 80 %.

32. The ideal efficiency for a heat engine operating between                (c)CM
    the temperatures of 227 degrees C and 27 degrees C is

        a. 20 %.               d. 88 %.
        b. 25 %.               e. none of these.
        c. 40 %.

# 17
## Vibrations and Waves

1. The source of all wave motion is a                                      (c)A

    a. wave pattern.        d. variable high and low
    b. parmonic object.       pressure regions.
    c. vibrating object.      e. none of these.

2. Unlike a transverse wave, a longitudinal wave has no                    (e)A

    a. amplitude.          d. speed.
    b. frequency.         e. ...a longitudinal wave
    c. wavelength.         has all of these.

3. In a longitudinal wave                                                  (a)A

    a. the compression and rarefactions travel in the same
       direction.
    b. the compressions and rarefactions travel in opposite
       directions.
    c. neither... only transverse waves have compressions
       and rarefactions.

4. An observer on the ground hears a sonic boom which is                   (d)A
created by an airplane flying at a speed

    a. just below the speed of sound.
    b. equal to the speed of sound.
    c. barely above the speed of sound.
    d. greater than the speed of sound.
    e. none of these.

5. An aircraft that flies faster than the speed of sound is                (b)A
said to be

    a. subsonic.        c. neither of these.
    b. supersonic.

6. A doppler effect occurs when a source of sound moves                    (c)A

    a. towards you.     c. both of the above.
    b. away from you.   d. none of the above.

7. A water molecule oscillates up and down two complete cycles             (c)AM
each second as a water wave passes by.  What is the wave's
frequency?

    a. 0.5 hertz.      d. 3 hertz.
    b. 1 hertz.        e. 6 hertz.
    c. 2 hertz.

8. A water molecule oscillates up and down two compete cycles    (c)AM
   in one second as a water wave passes by.  The wave's
   wavelength is 10 meters.  What is the wave's speed?

       a. 2 m/s.            d. 40 m/s.
       b. 10 m/s.           e. more than 40 m/s.
       c. 20 m/s.

9. A wave travels an average distance of 6 meters in one         (d)AM
   second. What is the wave's velocity?

       a. less than 0.2 m/s.    d. 6 m/s.
       b. 1 m/s.                e. more than 6 m/s.
       c. 3 m/s.

10. The frequency of the second hand on a clock is               (b)B

       a. 1 hertz.          c. 60 hertz.
       b. 1/60 hertz.

11. A doppler effect occurs when a source of sound moves         (a)B

       a. toward you.
       b. at right angles to you.
       c. both of the above.
       d. none of the above.

12. A weight on an end of a string bobs up and down one          (a)B
    complete cycle each two seconds.  Its frequency is

       a. 0.5 hertz.        c. neither of these.
       b. 2 hertz.

13. A weight on the end of a string bobs up and down one         (b)B
    complete cycle every two seconds.  Its period is

       a. 0.5 sec.          c. neither of these.
       b. 2 sec.

14. A weight suspended from a spring bobs up and down over a     (a)B
    distance of 1 meter in two seconds.  Its frequency is

       a. 0.5 hertz.        c. 2 hertz.
       b. 1 hertz.          d. none of these.

15. Some of a wave's energy is always being dissipated as heat.  (c)B
    In time, this will reduce the wave's

       a. speed.            d. frequency.
       b. wavelength.       e. period.
       c. amplitude.

16. The amplitude of a particular wave is 1 meter. The top to    (c)B
    bottom distance of the disturbance is

       a. 0.5 m.            c. 2 m.
       b. 1 m.              d. none of these.

17. When a pendulum clock at sea level is taken to the top of a  (b)B
    high mountain, it will

       a. gain time.        c. neither lose nor gain time.
       b. lose time.

18. If you double the frequency of a vibrating object, its (b)B
    period

        a. doubles.            c. is quartered.
        b. halves.

19. You dip your finger repeatedly into water and make waves. (a)B
    If you dip your finger more frequently, the wavelength of
    the waves

        a. shortens.           c. stays the same.
        b. lengthens.

20. During a single period, the distance traveled by a wave is (b)B

        a. one-half wavelength.
        b. one wavelength.
        c. two wavelengths.

21. A wave oscillates up and down two complete cycles each (d)BM
    second. If the wave travels an average distance of 6 meters
    in one second, its wavelength is

        a. 0.5 m.              d. 3 m.
        b. 1 m.                e. 6 m.
        c. 2 m.

22. Radio waves travels at the speed of light, 300,000 km/s. (b)BM
    The wavelength of a radio wave received at 100 megahertz is

        a. 0.33 m.             c. 3.3 m.
        b. 3.0 m.              d. 30 m.

23. As a train of water waves goes by, a cork on the water bobs (d)BM
    up and down one complete cycle each second.  The waves are
    2 meters long.  What is the speed of the wave?

        a. 0.25 m/s.           d. 2 m/s.
        b. 0.50 m/s.           e. 4 m/s.
        c. 1.0 m/s.

24. A skipper on a boat notices wave crests passing the anchor (a)BM
    chain every 5 seconds.  The skipper estimates the distance
    between crests is 15 m.  What is the speed of the water
    waves?

        a. 3 m/s.              c. 15 m/s.
        b. 5 m/s.              d. not enough information given.

25. A child swings back and forth on a playground swing.  If (b)C
    the child stands rather than sits, the time for a to-and-
    fro swing is

        a. lengthened.         c. unchanged.
        b. shortened.

26. Suppose a simple pendulum is suspended in an elevator. (b)C
    When the elevator is accelerating upward, the period of the
    pendulum

        a. increases.          c. doesn't change.
        b. decreases.

27. Horses would be able to run faster if most of the mass in                    (a)C
    their legs were concentrated

        a. in the upper part, nearer the horses' bodies.
        b. towards their feet.
        c. halfway up their legs.
        d. uniformly all along their legs.
        e. none of these.


28. The sonic boom produced by an aircraft flying at ground                       (d)C
    level will be reduced if the aircraft

        a. is smaller.          d. all of the above.
        b. flies higher.        e. none of the above.
        c. is more streamlined.


29. A tuning fork of frequency 200 hertz will resonate if a                        (a)C
    sound wave incident on it has a frequency of

        a. 100 hertz.           c. either of the above.
        b. 400 hertz.           d. none of the above.


30. If at a concert, a wind blows directly from the orchestra                      (c)C
    toward you, the frequency of the sound you hear will be

        a. decreased.           c. neither decreased
        b. increased.              nor increased.


31. If at a concert, a wind blows directly from the orchestra                      (b)C
    toward you, the speed of the sound you hear will be

        a. decreased.           c. neither decreased
        b. increased.              nor increased.


32. If at a concert, a wind blows directly from the orchestra                      (b)C
    toward you, the wavelength of the sound you hear will be

        a. decreased.           c. neither decreased
        b. increased.              nor increased.

# 18

## Sound

1. A sound wave is a                                                    (a)A

     a. longitudinal wave.   d. shock wave.
     b. transverse wave.    e. none of these.
     c. standing wave.

2. Sound waves cannot travel in                                        (d)A

     a. air.             d. a vacuum.
     b. water.          e. ... sound can travel in
     c. steel.            all of these.

3. The speed of a sound wave depends on                                 (c)A

     a. its frequency.    d. all of the above.
     b. its wavelength.   e. none of the above.
     c. the air temperature.

4. Sound travels faster in air if the air is                           (a)A

     a. warm.          c. neither warm nor cold.
     b. cold.

5. A wave having a frequency of 1000 hertz vibrates at                  (b)A

     a. less than 1000 cycles per second.
     b. 1000 cycles per second.
     c. more than 1000 cycles per second.

6. The phenomenon of beats results from sound                          (c)A

     a. refraction.      d. all of the above.
     b. reflection.      e. none of the above.
     c. interference.

7. Caruso is said to have made a crystal chandelier shatter            (d)A
  with his voice.  This is a demonstration of

     a. an echo.        d. resonance.
     b. sound refraction.  e. interference.
     c. beats.

8. Sound waves can interfere with one another so that no sound         (a)A
  results.

     a. true.          b. false.

9. When a source of sound moves away from you, you hear a decrease in the sound's

   a. frequency.          d. all of the above.
   b. speed.              e. none of the above.
   c. wavelength.

(a)A

10. When a source of sound approaches you, you hear an increase in the sound's

    a. frequency.         d. all of the above.
    b. speed.             e. none of the above.
    c. wavelength.

(a)A

11. Sound interference is the reason for

    a. resonance.         d. echoes.
    b. beats.             e. the Doppler effect.
    c. refraction.

(b)A

12. On some days, air nearest the ground is colder than air that is higher up.  On one of these days, sound waves

    a. tend to be refracted upward.
    b. tend to be refracted downward.
    c. travel without refraction.

(b)B

13. Sound refraction depends on the fact that the speed of sound is

    a. constant.
    b. variable.
    c. proportional to frequency.
    d. inversely proportional to wavelength.
    e. none of these.

(b)B

14. A 340 hertz sound wave travels at 340 m/s in air with a wavelength of

    a. 1 m.               d. 1000 m.
    b. 10 m.              e. none of these.
    c. 100 m.

(a)B

15. When the handle of a tuning fork is held solidly against a table, the sound becomes louder and the time the fork keeps vibrating becomes

    a. longer.            c. ... remains the same.
    b. shorter.

(b)B

16. Beats are produced when two tuning forks, one of frequency 240 hertz and the other of frequency 246 hertz are sounded together.  The frequency of the beats is

    a. 6 hertz.           d. 245 hertz.
    b. 12 hertz.          e. none of these.
    c. 240 hertz.

(a)B

17. In which one of these does sound travel the fastest?

    a. water vapor.       c. steam.
    b. water.             d. ...sound travels the same
    c. ice.                   speed in each of these.

(c)B

18. A 1056 hertz turning fork is sounded at the same time a    (d)B
    piano note is struck.  You hear two beats per second.  What
    is the frequency of the piano string?

        a. 1053 hertz.          c. 2112 hertz.
        b. 1056 hertz.          d. not enough information
        c. 1059 hertz.             given to determine.

19. Inhaling helium increases the pitch of your voice.  The    (b)B
    reason for this is, compared to air, sound travels

        a. slower in helium.
        b. faster in helium.
        c. the same speed in helium, but the wavelength is more.

20. An explosion occurs 34 km away.  Knowing that sound travels at    (e)BM
    340 m/s, the time it takes for the sound to reach you is

        a. 0.1 sec.             d. 20 sec.
        b. 1 sec.               e. more than 20 sec.
        c. 10 sec.

21. Xenon has atomic number 54, while Krypton has atomic    (b)C
    number 36. Given that both gases are at the same
    temperature, in which medium does sound travel fastest?

        a. Xenon gas.           c. a mixture of both gases.
        b. Krypton gas.         d. ...the same in either gas.

22. A neon atom has a larger atomic number than a helium atom.    (b)C
    How does the speed of sound in helium gas compare to the
    speed of sound in neon gas given the same temperature
    and pressure?

        a. greater in neon than in helium.
        b. greater in helium than in neon.
        c. the same in both gases.
        d. depends on the frequency of the sound generated.

23. A general rule for estimating the distance in kilometers    (b)CM
    between an observer and a lightning bolt is to count the
    number of seconds between seeing the lightning and hearing
    it, and dividing by

        a. 2.                   d. 5.
        b. 3.                   e. none of these.
        c. 4.

24. Two whistles produce sounds of wavelengths 3.4 m and 3.3 m.    (d)CM
    What is the beat frequency produced?

        a. 0.1 hertz.           d. 3.0 hertz.
        b. 1.0 hertz.           e. 4.0 hertz.
        c. 2.0 hertz.

25. Suppose you sound a tuning fork at the same time you hit a    (b)CM
    1056 hertz note on the piano and hear 2 beats/sec.  You
    tighten the piano string very slightly and now hear 3
    beats/sec.  What is the frequency of the tuning fork?

        a. 1053 hertz.          d. 1058 hertz.
        b. 1054 hertz.          e. 1059 hertz.
        c. 1056 hertz.

# 19

# Musical Sounds

Chapter 19: Music

1. The pitch of a musical sound depends on the sound wave's          (b)A

    a. wavelength.         d. amplitude.
    b. frequency.         e. all of these.
    c. speed.

2. The loudness of a musical sound is a measure of the          (d)A
sound wave's

    a. wavelength.         d. amplitude.
    b. frequency.         e. all of these.
    c. speed.

3. A decibel is a measure of a sound's          (d)A

    a. frequency.         d. loudness.
    b. wavelength.       e. all of these.
    b. speed.

4. The quality of a musical note has to do with its          (c)A

    a. loudness.         d. amplitude.
    b. frequency.         e. all of these.
    c. overtones.

5. High pitched sound has a          (b)A

    a. high speed.       d. all of the above.
    b. high frequency.   e. none of the above.
    c. long wavelength.

6. Reverberation is a phenomenon you would be most likely to          (a)A
hear if you sing in the

    a. shower.         b. desert.

7. The fundamental frequency of a violin string is 440 hertz.          (c)A
The frequency of its first overtone is

    a. 220 hertz.       c. 880 hertz.
    b. 440 hertz.       d. none of these.

8. All other things being the same, strings that have more          (b)B
mass than other strings will have

    a. higher frequency notes.
    b. lower frequency notes.
    c. the same frequency notes.

9. If a guitar has no sounding board, a note played on the guitar will sound for a    (c)B

    a. shorter time.          c. longer time.
    b. the same length of time.


10. What is the threshhold of hearing?    (e)A

    a. 10 decibel.            d. 0.01 decibel.
    b. 1 decibel.             e. none of these.
    c. 0.1 decibel.


11. Compared to a sound of 40 decibels, a sound of 50 decibels is    (a)B

    a. 10 times louder.       d. 1000 times louder.
    b. 100 times louder.      e. more than 1000 times louder.
    c. 1000 times louder.


12. About how many octaves are present between 20 hertz and 2560 hertz?    (c)CM

    a. 5.                     d. 8.
    b. 6.                     e. 9.
    c. 7.


13. About how many octaves are present between 100 hertz and 1600 hertz?    (a)CM

    a. 4.                     d. 7.
    b. 5.                     e. 8.
    c. 6.


14. What is the wavelength of the third overtone of a 340 hertz note?  (The speed of sound is 340 m/s.)    (b)CM

    a. 0.25 m.                d. 1.0 m.
    b. 0.33 m.                e. none of these.
    c. 0.5 m.


15. A cello string 0.75 m long has a fundamental frequency of 220 hertz.  The speed of a wave on the string is    (d)CM

    a. 165 m/s.               d. 330 m/s.
    b. 220 m/s.               e. none of these.
    c. 294 m/s.


16. The wave speed on a tightened guitar string is 880 m/s. What is shortest length of string that will produce standing waves of 440 hertz frequency?    (b)CM

    a. 0.5 m.                 d. 2.0 m.
    b. 1.0 m.                 e. none of these.
    c. 1.5 m.

# 20

# Electrostatics

1. Two like charges                                                      (b)A

   a. attract each other.    d. have no effect on each other.
   b. repel each other.      e. must be neutrons.
   c. neutralize each other.

2. Protons and electrons                                                 (b)A

   a. repel each other.      c. do not interact.
   b. attract each other.

3. Which force binds atoms together to form molecules?                   (c)A

   a. gravitational.         d. centripetal.
   b. nuclear.               e. none of these.
   c. electrical.

4. The fundamental force underlying all chemical reactions is            (d)A

   a. gravitational.         d. electrical.
   b. nuclear.               e. none of these.
   c. centripetal.

5. The electrical force between charges is strongest when the            (a)A
   charges are

   a. close together.        c. ...the electric force is
   b. far apart.                 constant everywhere.

6. The electrical force between charges depends on                       (c)A

   a. how large the charges are.
   b. how far apart the charges are.
   c. both of the above.
   d. none of these.

7. When the distance between two charges is halved, the                  (a)B
   electrical force between the charges

   a. quadruples.            d. is reduced by 1/4.
   b. doubles.               e. none of these.
   c. halves.

8. Particle A has twice the charge as particle B.  Compared              (c)B
   to the force on the particle A, the force on the particle B
   is

   a. four times as much.    d. half as much.
   b. two times as much.     e. none of these.
   c. the same.

9. If you comb your hair and the comb becomes positively    (b)B
   charged, then your hair becomes

        a. positively charged.   c. uncharged.
        b. negatively charged.

10. To say that electric charge is conserved means that no case    (d)B
    has ever been found where

        a. the total charge on an object has changed.
        b. the net amount of negative charge on an object is
           unbalanced by a positive charge on another object.
        c. the total amount of charge on an object has increased.
        d. net charge has been created or destroyed.
        e. none of these.

11. A difference between electric forces and gravitational forces    (b)B
    is that electrical forces include

        a. separation distance.
        b. repulsive interactions.
        c. the inverse square law.
        d. infinite range.
        e. none of these.

12. A conductor differs from an insulator in that a conductor    (e)B

        a. has more electrons than protons.
        b. has more protons than electrons.
        c. has more energy than an insulator.
        d. has faster moving molecules.
        e. none of these.

13. A negatively charged rod is brought near a metal can that    (a)B
    rests on a wood table. You touch the opposite side of the
    can momentarily with your finger. The can is then

        a. positively charged.   c. uncharged.
        b. negatively charged.   d. charged the same as it was.

14. Every proton in the universe is surrounded by its own    (c)B

        a. electric field.       c. both of the above.
        b. gravitational field.  d. none of the above.

15. The electric field around an isolated electron has a certain    (e)B
    strength 1 cm from the electron. The electric field strength
    2 cm from the electron is

        a. half as much.         d. four times as much.
        b. the same.             e. none of these.
        c. twice as much.

16. If you use 10 J of work to push a unit charge into an    (b)B
    electric field, its voltage with respect to its starting
    position is

        a. less than 10 V.       c. more then 10 V.
        b. 10 V.

17. If you use 10 J of work to push a charge into an electric    (b)B
    field and then release the charge, as it flies past its
    starting position, its kinetic energy is

        a. less than 10 J.       c. more than 10 J.
        b. 10 J.

18. An electroscope is charged positively as shown by foil          (a)B
    leaves that stand apart.  As a negative charge is brought
    close to the electroscope, the leaves

        a. fall closer together.
        b. spread apart further.
        c. do not move.

19. Charge carriers in a metal are electrons rather than protons    (d)B
    because electrons are

        a. loosely bound.       d. all of the above.
        b. lighter.             e. none of the above.
        c. far from a nucleus.

20. To be safe in the unlikely case of a lightning strike, it       (a)B
    is best to be inside a buiding framed with

        a. steel.               c. both the same.
        b. wood.

21. Two charges separated by one meter exert a one N force on        (e)BM
    each other.  If the charges are pushed to 1/4 meter
    separation, the force on each charge will be

        a. 1 N.                 d. 8 N.
        b. 2 N.                 e. 16 N.
        c. 4 N.

22. Two charges separated by one meter exert a one N force on        (b)BM
    each other.  If the charges are pulled to 3 m separation
    distance, the force on each charge will be

        a. 0.33 N.              d. 3 N.
        b. 0.11 N.              e. 9 N.
        c. 0 N.

23. Two charges separated by one meter exert a one N force on        (c)BM
    each other.  If the magnitude of each charge is doubled, the
    force on each charge is

        a. 1 N.                 d. 8 N.
        b. 2 N.                 e. none of these.
        c. 4 N.

24. The electrical force on a 2 C charge is 60 N.  What is the       (b)BM
    value of the electric field at the place where the charge
    is located?

        a. 20 N/C.              d. 120 N/C.
        b. 30 N/C.              e. 240 N/C.
        c. 60 N/C.

25. An electron is pushed into an electric field where it           (c)C
    acquires a 1 V electrical potential.  If two electrons are
    pushed the same distance into the same electric field, the
    electrical potential of the two electrons is

        a. 0.25 V.              d. 2 V.
        b. 0.5 V.               e. 4 V.
        c. 1 V.

26. The electric field inside an uncharged metal ball is zero.      (b)C
    If the ball is negatively charged, the electric field inside
    the ball is then

        a. less than zero.      c. greater than zero.
        b. zero.

- 119 -

27. Two charged particles held close to each other are          (b)C
released.  As they move, the force on each particle
increases. Therefore, the particles have

    a. the same sign.    c. not enough information
    b. the opposite sign.      given.

28. Two charged particle held close to each other are released.   (c)C
As the particles move, the velocity of each increases.
Therefore, the particles have

    a. the same sign.    c. not enough information
    b. the opposite sign.      given.

29. A positive charge and a negative charge held near each other   (a)C
are released.  As they move, the force on each particle

    a. increases.    c. stays the same.
    b. decreases.

30. Two charged particles held near each other are released. As    (a)C
they move, the acceleration of each decreases.  Therefore,
the particles have

    a. the same sign.    c. not enough information
    b. the opposite sign.      given.

# 21

# Electric Current

1. Electrons move in an electrical circuit                              (c)A

    a. by being bumped by other electrons.
    b. by colliding with molecules.
    c. by interacting with an established electric field.
    d. because the wires are so thin.
    e. none of these.


2. In an electrical circuit, electrons come from                        (d)A

    a. a dry cell, wet cell or battery.
    b. the back emf of motors.
    c. the power station generator.
    d. the electrical conductor itself.
    e. none of these.


3. The source of electrons lighting an incandescent light bulb is       (c)A

    a. the power company.     d. the wire leading to the lamp.
    b. electrical outlet.     e. the source voltage.
    c. atoms in the light bulb filament.


4. A woman experiences an electrical shock.  The electrons              (a)A
   making the shock come from

    a. the woman's body.      b. the object causing the shock.
    b. the ground.            e. electric field in the air.
    c. a nearby power supply.

5. In a common circuit, electrons move ~~at~~ *along the circuit* speeds of    (a)A

    a. a fraction of a centimeter per second.
    b. many centimeters per second.
    c. the speed of a sound wave.
    d. the speed of light.
    e. none of these.


6. When a light switch is turned on in a DC circuit, the               (d)A
   average speed of electrons in the lamp is

    a. the speed of sound waves in metal.
    b. the speed of light.
    c. 1000 cm/s.
    d. less than 1 cm/s.
    e. dependent on how quickly each electron bumps into the
       next electron.


7. A wire carrying a current is normally charged                        (c)A

    a. negatively.            c. not at all.
    b. positively.

8. In an AC circuit, the electric field                                    (b)A

    a. increases via the inverse square law.
    b. changes magnitude and direction with time.
    b. is everywhere the same.
    d. is non-existent.
    e. none of these.


9. The current through a 10 ohm resistor connected to a 120 V       (c)AM
    power supply is

    a. 1 A.                    d. 120 A.
    b. 10 A.                   e. none of these.
    c. 12 A.


10. A 10 ohm resistor has 5 A current in it.  What is the          (e)AM
    voltage across the resistor?

    a. 5 V.                    d. 20 V.
    b. 10 V.                   e. more than 20 V.
    c. 15 V.


11. When a 10 V battery is connected to a resistor, 2 A of         (b)AM
    current flow in the resistor.  What is the resistor's value?

    a. 2 ohms.                 d. 20 ohms.
    b. 5 ohms.                 e. more than 20 ohms.
    c. 10 ohms.


12. The number of electrons delivered to an average American home   (a)B
    by an average power utility in 1984 was

    a. zero.                   d. billions of billions.
    b. 110.                    e. none of these.
    c. 220.


13. Two lamps, one with a thick filament and one with a thin        (c)B
    filament, are connected in series.  The current is greater
    in the lamp with the

    a. thick filament.         c. ...the same in each
    b. thin filament.          lamp.


14. Two lamps, one with a thick filament and one with a thin        (c)B
    filament, are connected in parallel to a battery.  The volt-
    age in greatest across the lamp with the

    a. thick filament.         c. ...both voltages are
    b. thin filament.          the same.


15. Two lamps, one with a thick filament and one with a thin        (a)B
    filament, are connected in parallel to a battery.  The
    current is largest in the lamp with the

    a. thick filament.         c. ... current is the same in both.
    b. thin filament.


16. Two lamps, one with a thick filament and one with a thin        (b)B
    filament are connected in series to a battery.  The voltage
    is greater across the lamp with the

    a. thick filament.         c. ...voltage is the same for both.
    b. thin filament.

17. As more lamps are put into a series circuit, the overall          (b)B
    current in the circuit

        a. increases.          c. stays the same.
        b. decreases.

18. As more lamps are put into a parallel circuit, the overall        (a)B
    current in the circuit

        a. increases.          c. stays the same.
        b. decreases.

19. A circuit is powered with a battery.  Current flows               (d)B

        a. out of the battery and into the circuit.
        b. from the negative battery terminal to the positive
           terminal.
        c. after a couple seconds passes.
        d. through both the battery and rest of the circuit.
        e. none of these.

20. When we say an appliance "uses up" electricity, we really         (e)B
    are saying that

        a. current disappears.
        b. electric charges are dissipated.
        c. the main power supply voltage is lowered.
        d. electrons are taken out of the circuit and put some-
           where else.
        e. electron kinetic energy is changed into heat.

21. Compared to the resistance of two lamps connected in series,      (b)B
    the same two lamps connected in parallel have

        a. more resistance.          c. the same resistance.
        b. less resistance.

22. If you plug an electric toaster rated at 110 V into a 220 V       (d)B
    outlet, current in the toaster will be

        a. half what is should be.
        b. the same as if it were plugged into 110 V.
        c. more than twice what it should be.
        d. twice what it should be.
        e. none of these.

23. A 60 Watt light bulb is connected to a 120 volt plug.  What       (b)BM
    is the current in the light bulb?

        a. 0.25 A.          d. 4 A.
        b. 0.5 A.           e. more than 4 A.
        c. 2 A.

24. A light bulb is plugged into a 120 Volt outlet and has 0.8 A      (d)BM
    current in it.  What is the power rating of the light bulb?

        a. 12 W.           d. 96 W.
        b. 15 W.           e. 120 W.
        c. 60 W.

25. A 60 W light bulb and a 100 W light bulb are each rated at        (a)C
    120 V.  Which light bulb has a larger resistance?

        a. the 60 W bulb.          c. ...both have the same
        b. the 100 W bulb.            resistance.

- 123 -

26. A 60 W light bulb and a 100 W light bulb are each connected     (b)C
    to a 120 V outlet.  Which light bulb has more current in
    it?

        a. the 60 W bulb.        c. ...both have the same
        b. the 100 W bulb.          current.

27. Compared to the filament thickness on a 60 W light bulb,        (b)C
    the filament thickness of a 100 W light bulb will be

        a. less.                 c. the same.
        b. greater.

28. Compared to a single lamp connected to a battery, two           (b)C
    identical lamps connected in series to the same battery
    will use

        a. more current.         c. the same current.
        b. less current.

29. Compared to a single lamp connected to a battery, two lamps     (a)C
    connected in parallel to the same battery will use

        a. more current.         c. the same current.
        b. less current.

30. An electric heater is rated at 300 W when used in a 110 V       (d)CM
    circuit.  The safety fuse in the circuit can handle 15 A
    of current.  How many heaters can be safely operated in the
    circuit?

        a. 2.                    d. 5.
        b. 3.                    e. more than 5.
        c. 4.

31. A heater uses 20 A when used in a 110 V line.  If electric      (c)CM
    power costs 10 cents per kilowatt hour, the cost of running
    the heater for 10 hours is

        a. $0.22.                d. $5.50.
        b. $0.55.                e. none of these.
        c. $2.20.

32. The current through two identical light bulbs connected in      (c)CM
    series is 0.25 A.  The voltage across both bulbs is 110 V.
    The resistance of a single light bulb is

        a. 22 ohms.              d. 440 ohms.
        b. 44 ohms.              e. none of these.
        c. 220 ohms.

33. A power line with a resistance of 2 ohms has a current of       (d)CM
    80 A current in it.  The power dissipated in the line is

        a. 40 W.                 d. 12800 W.
        b. 160 W.                e. none of these.
        c. 320 W.

34. What is the resistance of a 120 W incandescent lamp connected   (e)CM
    to a 120 V power supply?

        a. 1 ohm.                d. 144 ohms.
        b. 60 ohms.              e. none of these.
        c. 100 ohms.

# 22

# Magnetism

1. The source of all magnetism is                                                        (d)A

    a. tiny pieces of iron.
    b. tiny domains of aligned atoms.
    c. ferromagnetic materials.
    d. moving electric charge.
    e. none of these.


2. Moving electric charges will interact with                                            (c)A

    a. an electric field.     c. both of the above.
    b. a magnetic field.      d. none of the above.


3. An iron rod becomes magnetic when                                                     (c)A

    a. positive ions accumulate at one end and negative
       ions at the other end.
    b. its atoms are aligned having plus charges on one side
       and negative charges on the other.
    c. the net spins of its electrons are in the same
       direction.
    d. its electrons stop moving and point in the same
       direction.
    e. none of these.


4. Surrounding every moving electron is                                                  (c)A

    a. a magnetic field.      c. both of the above.
    b. an electric field.     d. none of the above.


5. Magnetism is due to the motion of electrons as they                                   (c)A

    a. move around the nucleus.
    b. spin on their axes.
    c. both of the above.
    d. none of the above.


6. The force on an electron moving in a magnetic field will be                           (c)A
   the largest when its direction

    a. is the same as the magnetic field direction.
    b. is exactly opposite to the magnetic field direction.
    c. is perpendicular to the magnetic field direction.
    d. is at an angle other than 90 degrees to the magnetic
       field direction.
    e. none of these.


7. The intensity of cosmic rays bombarding the Earth's surface                           (a)A
   is largest at the

    a. poles.                 c. equator.
    b. mid-latitudes.

8. Which geographic pole of the Earth is nearest the magnetic north pole of the Earth?　　　　　　　　　　　　　　　　(b)A

    a. north pole.　　　　　c. both of these.
    b. south pole.

9. Which pole of a compass needle points to a south pole of a magnet?　　　　　　　　　　　　　　　　　　　　　　　(a)A

    a. north pole.　　　　　c. both of these.
    b. south pole.

10. Pigeons navigate primarily by　　　　　　　　　　　　(c)A

    a. a good memory.
    b. a keen sense of smell.
    c. a magnetic sensor in their heads.
    d. ultra-high pitched sounds.
    e. none of these.

11. If a compass is moved from the northern hemisphere to the southern hemisphere, its magnetic needle will　　　　　(d)B

    a. change direction by 180 degrees.
    b. change direction depending on where the measurement
       is taken.
    c. change direction by 90 degrees.
    d. not change direction at all.
    e. none of these

12. The magnetic poles of the Earth are located　　　　(c)B

    a. at the Earth's geographic poles.
    b. a few meters from the Earth's geographic poles.
    c. thousands of kilometers from the Earth's geographic
       poles.
    d. closer to the equator than the geographic poles.
    e. ...nowhere, the Earth doesn't have a magnetic field.

13. A possible cause for the existence of the Earth's magnetic field is　　　　　　　　　　　　　　　　　　　　　(d)B

    a. moving charges in the liquid part of the Earth's core.
    b. great numbers of very slow moving charges in the Earth.
    c. convection currents in the liquid part of the Earth's
       core.
    d. all of the above.
    e. none of the above.

14. Outside a magnet, magnetic field lines are conventionally drawn from　　　　　　　　　　　　　　　　　　　　　(a)B

    a. north to south.　　　c. either way.
    b. south to north.

15. Which force field can increase a moving electron's speed?　　(a)C

    a. electric field.　　　c. both of the above.
    b. magnetic field.　　　d. none of the above.

16. Which force field can accelerate an electron, but never change its speed?　　　　　　　　　　　　　　　　　　(b)C

    a. electric field.　　　c. both of the above.
    b. magnetic field.　　　d. none of the above.

# 23

# Electromagnetic Interactions

1. The magnetic field strength inside a current carrying coil      (d)A
will be greater if the coil encloses a

    a. vacuum.             d. iron rod.
    b. wooden rod.       e. none of these.
    c. glass rod.

2. A device that transforms electrical energy to mechanical      (b)A
energy is a

    a. generator.          d. magnet.
    b. motor.             e. none of these.
    c. transformer.

3. If a magnet is pushed into a coil, voltage is induced across      (c)A
the coil. If the same magnet is pushed into a coil with
twice the number of loops,

    a. one half as much voltage is induced.
    b. the same voltage is induced.
    c. twice as much voltage is induced.
    d. four times as much voltage is induced.
    e. none of these.

4. A step-up transformer increases      (d)A

    a. power.           c. both of the above.
    b. energy.          d. none of the above.

5. The principal advantage of AC power over DC power is that      (c)B

    a. less energy is dissipated during transmission.
    b. AC voltage oscillates while DC voltage does not.
    c. AC voltage can be tranformed more easily.
    d. AC circuits multiply power more easily.
    e. AC circuits are safer.

6. The major advantage of MHD generators over conventional      (e)B
generators is that MHD generators

    a. don't use electromagnetic induction.
    b. can be powered with superconducting magnets.
    c. require no power input.
    d. all of the above.
    e. none of the above.

7. Neon signs require about 12,000 volts to operate. If the      (a)D
circuit uses a 120 volt power source, the ratio of primary
to secondary turns on the transformer should be

    a. 1:100.          c. neither of these.
    b. 100:1.

8. A step-up transformer has a ratio of one to ten.  If 100 W          (c)B
   of power go into the primary coil, the power coming from the
   secondary coil is

      a. 1 W.               d. 1000 W.
      b. 10 W.             e. none of these.
      c. 100 W.

9. The voltage across the input terminals of a transformer is       (b)BM
   110 V.  The primary has 50 loops and the secondary has 25
   loops.  The voltage the transformer puts out is

      a. 25 V.             d. 220 V.
      b. 55 V.             e. none of these.
      c. 110 V.

10. Four amps of current exist in the primary coil of a             (c)BM
    transformer.  The voltage across the primary coil is 110 V.
    What is the power output of the secondary coil?

      a. less than 30 W.
      b. 110 W.        e. not enough information to say.
      c. 440 W.

11. A certain transformer doubles input voltage.  If the primary    (a)BM
    coil has 10 A of current, then the current in the secondary
    coil is

      a. 5 A.              d. 20 A.
      b. 10 A.             e. none of these.
      c. 20 A.

12. Technically, an automobile will use more gas if the lights      (b)C
    are turned on.  This statement is

      a. true only if the engine is running.
      b. true whether or not the engine is running.
      c. not true.

13. As a motor armature turns faster and faster, net current in     (b)C
    the motor windings

      a. increases.       c. remains unchanged.
      b. decreases.

14. If the primary of a transformer were connected to a DC power    (d)C
    source, the transformer would operate

      a. at very low efficiency.
      b. the same as always.
      c. only if the output is also DC voltage.
      d. only while being connected or disconnected.
      e. none of these.

# 24

# Electromagnetic Radiation

1. The fact that light travels at exactly 300,000 km/s is a consequence of

   (d)A

   a. electromagnetic wave propagation.
   b. electromagnetic field induction.
   c. Maxwell's laws.
   d. energy conservation.
   e. none of these.

2. The source of all electromagnetic waves is

   (c)A

   a. changes in atomic energy levels.
   b. vibrating atoms.
   c. accelerating charges.
   d. crystalline fluctuations.
   e. none of these.

3. Which type of radio wave has the higher frequency?

   (b)A

   a. AM.                    c. both have the same
   b. FM.                       frequency.

4. Which type of radio wave has the least static when used?

   (b)A

   a. AM.                    c. both have equal amounts
   b. FM.                       of static.

5. Electromagnetic waves can

   (a)A

   a. travel through a vacuum.
   b. need a medium to travel through.

6. Which of these electromagnetic waves has the shortest wavelength?

   (c)A

   a. radio waves.        d. ultraviolet waves.
   b. infrared waves.     e. light waves.
   c. X-rays.

7. Compared to ultraviolet waves, the wavelength of infrared waves is

   (b)A

   a. shorter.            c. the same.
   b. longer.

8. Compared to radio waves, the velocity of visible light waves is

   (c)A

   a. less.               c. the same.
   b. more.

9. If an electric charge is shaken up and down,　　　　　　　　　　(d)A

    a. sound is emitted.　　d. a magnetic field is created.
    b. light is emitted.　　e. its mass decreases.
    c. electron excitation occurs.


10. If an electron vibrates up and down 1000 times each second,　　(d)A
    it generates an electromagnetic wave having a

    a. period of 1000 s.　　d. frequency of 1000 Hz.
    b. speed of 1000 m/s.　e. wavelength of 1000 m.
    c. wavelength of 1000 m.


11. The source of all electromagnetic waves is　　　　　　　　　　(d)A

    a. heat.　　　　　　　d. vibrating charges.
    b. magnetic fields.　e. none of these.
    c. electric fields.


12. Which of the following is fundamentally different from the　　(a)A
    others?

    a. sound waves.　　　d. light waves.
    b. X-rays.　　　　　　e. radio waves.
    c. gamma rays.


13. To say a wave is modulated means that the wave is　　　　　　(a)A

    a. changed.　　　　　b. averaged.
    b. smoothed.　　　　　e. amplified.
    c. diminished.


14. A radio wave has the same frequency as　　　　　　　　　　　(d)B

    a. an X-ray.
    b. blue light waves.
    c. the sound it produces.
    d. electrons in the transmitting antenna.
    e. none of these.


15. A certain object emits infrared waves. If it were to emit　　(a)B
    light waves instead, its temperature would have to be

    a. higher.　　　　　　c. the same, temperature doesn't
    b. lower.　　　　　　　　 make any difference.


16. The main difference between a radio wave and a light wave　　(b)B
    is its

    a. speed.　　　　　　　c. both of the above.
    b. wavelength.　　　　d. none of the above.


17. The main difference between a radio wave and a sound wave is　(e)B
    its

    a. frequency.　　　　d. amplitude.
    b. wavelength.　　　　e. basic nature.
    c. energy.


18. Which of the following continually emits electromagnetic　　(d)B
    radiation?

    a. insects.　　　　　　d. all of the above.
    b. radio antennas.　　e. none of the above.
    c. red-hot coals.

19. A lamp filament from a broken light bulb is continually          (b)B

    a. increasing in temperature.
    b. emitting electromagnetic radiation.
    c. increasing in mass.
    d. drawing electricity.
    e. none of these.


20. If the sun were to disappear right now, we wouldn't know         (c)B
    about it for 8 minutes because it takes

    a. 8 minutes for the sun to disappear.
    b. 8 minutes to operate receiving equipment in the dark.
    c. 8 minutes for light to travel from the sun to the Earth.
    d. all of the above.
    e. none of the above.


21. What is the wavelength of an electromagnetic wave that has       (c)BM
    a 1 Hz frequency?

    a. less than 1 m.        c. more than 1 m.
    b. 1 m.


22. What is the wavelength of an electromagnetic wave having a       (c)BM
    3 kilohertz frequency?

    a. less than 1 km.       c. more than 1 km.
    b. 1 km.


23. What is the frequency of an electromagnetic wave having a        (b)BM
    wavelength of 300,000 km?

    a. less than 1 Hz.       c. more than 1 Hz.
    b. 1 Hz.

# 25

# Reflection and Refraction

1. Diffuse reflection occurs when the size of surface irregular-    (b)A
   ities is

   a. small compared to the wavelength of the light used.
   b. large compared to the wavelength of the light used.
   c. larger than 1 cm in diameter.
   d. larger than 1 m in diameter.
   e. none of these.

2. The speed of light is greatest in                               (a)A

   a. red glass.          d. blue glass.
   b. orange glass.       e. ...is the same in all
   c. green glass.           of these.

3. A beam of light emerges from water into air at an angle.  The   (c)A
   beam is bent

   a. towards the normal  d. 96 degrees upward.
   b. 48 degrees upward.  e. not at all.
   c. away from the normal.

4. When a light beam emerges from water into air, the average      (a)A
   light speed

   a. increases.          c. remains the same.
   b. decreases.

5. Refraction results from differences in light's                  (c)A

   a. frequency.          d. all of the above.
   b. incident angles.    e. none of the above.
   c. speed.

6. Light refracts when traveling from air into glass because       (e)A
   light

   a. intensity is greater in air than in glass.
   b. intensity is greater in glass than in air.
   c. frequency is greater in air than in glass.
   d. frequency is greater in glass than in air.
   e. travels slower in glass than in air.

7. The effect that we call a mirage has most to do with            (d)A

   a. scattering.         d. refraction.
   b. interference.       e. diffraction.
   c. reflection.

8. When light reflects from a surface, there is a change in its     (e)B

    a. frequency.       d. all of the above.
    b. wavelength.     e. none of the above.
    c. speed.

9. If a fish looks upward at 45 degrees with respect to the     (c)B
water's surface, it will see

    a. the bottom of the pond.
    b. another fish in the pond.
    c. the sky and possibly some hills.
    d. only the water's surface.
    e. none of these.

10. The reason diamonds looks many colored is because     (d)B

    a. diamonds are extremely hard.
    b. light cannot enter diamond.
    c. diamonds are forever.
    d. different colors of light travel at different speeds
       in diamonds.
    e. none of these.

11. Rainbows exist because light is     (c)B

    a. reflected.      c. both of the above.
    b. refracted.      d. none of the above.

12. A person standing waist-deep in a swimming pool appears to be     (e)B
have short legs because of light

    a. reflection.      d. diffraction.
    b. absorption.     e. refraction.
    c. interference.

13. Different colors of light travel at different speeds in a     (b)B
transparent medium.  In a vacuum, different colors of light
travel at

    a. different speeds.    c. ... light travels at the same
    b. the same speed.       speed everywhere.

14. When white light goes from air into water, the color that     (d)B
refracts the most is

    a. red.         d. violet.
    b. orange.      e. all refract the same amount.
    c. green.

15. A secondary rainbow is dimmer than a primary rainbow because     (b)B

    a. sunlight reaching it is less intense.
    b. there is an extra reflection inside the water drops.
    c. only large drops produce secondary rainbows.
    d. it is a reflection of the primary rainbow.
    e. none of these.

16. If you walk towards a mirror at a certain speed, the relative     (c)B
speed between you and your image is

    a. half your speed.    c. twice your speed.
    b. your speed.       d. none of these.

17. A beam of light travels fastest in                                    (d)B

       a. glass.          d. air.
       b. water.          e. ...is the same in each
       c. plastic.           of these.


18. When seen from an airplane, a rainbow sometimes forms a               (a)B
    complete circle.  When this happens, the plane's shadow is

       a. in the center of the rainbow.
       b. in the lower part of the rainbow.
       c. in the upper part of the rainbow.
       d. totally outside the rainbow.
       e. nowhere, there is no shadow.


19. A fish outside water will see better if it has goggles that           (c)B
    are

       a. tinted blue.        d. extremely shiny.
       b. hemispheres.        e. none of these.
       c. filled with water.


20. Which of the following are consequences of light traveling           (d)B
    at different speeds in different media?

       a. mirages.            d. all of these.
       b. rainbows.           e. none of these.
       c. internal reflection.


21. Stars twinkle when seen from the Earth.  When seen from the           (c)B
    Moon, stars

       a. twinkle more.       c. don't twinkle.
       b. twinkle less.


22. Atmospheric refraction tends to make daytimes                         (a)B

       a. longer.             c. ... no change in day length.
       b. shorter.


23. Fermat's principle of least time applies to                          (c)B

       a. reflection.         c. both of the above.
       b. refraction.         d. none of the above.


24. Fermat's principle of least time could as well be the                (a)B
    principle of least distance for the case of

       a. reflection.         c. both of the above.
       b. refraction.         d. none of the above.


25. The sun's elliptical shape at sunset can be adequately               (c)B
    explained by

       a. Fermat's principle.
       b. the law of refraction.
       c. both of the above.
       d. none of the above.


26. If you were underwater spearing a fish with a laser beam,            (d)B
    you should aim the beam

       a. above the apparent position of the fish.
       b. below the apparent position of the fish.
       c. either of the above.
       d. none of the above.

27. When light is refracted, there is a change in its                (b)C

      a. frequency.        c. both of the above.
      b. wavelength.     d. none of the above.

28. At the same time an astronaut on the Moon sees a solar            (a)C
    eclipse, observers on Earth see

      a. a lunar eclipse.    c. no eclipse at all.
      b. a solar eclipse.

29. The moon's redness during a lunar eclipse results from            (d)C

      a. only lower frequencies being reflected from the moon.
      b. infrared radiation that is normally blocked.
      c. an optical illusion.
      d. refraction from all the sunsets on Earth.
      e. none of these.

30. When a pulse of white light is incident on a piece of glass,      (a)C
    the first color to emerge is

      a. red.           d. violet.
      b. orange.       c. ...they all emerge at the
      c. green.          time.

31. The shortest plane mirror in which you can see your entire        (a)C
    image is

      a. half your height.
      b. twice your height.
      c. equal to your height.
      d. depends on how far the mirror is from you.
      e. can not be determined.

32. To see his full height, a boy that is 1 meter tall needs a        (b)C
    mirror that is at least

      a. 0.25 m tall.     d. 2 m tall.
      b. 0.50 m tall.    e. depends on how far from the
      c. 1 m tall.         mirror is from the boy.

33. Magnification from a lens would be greater if light              (c)C

      a. propagated instantaneously.
      b. traveled faster in glass than it does.
      c. traveled slower in glass than it does.
      d. beams spread out more.
      e. none of these.

34. When you look at youself in a pocket mirror, and then hold        (a)C
    the mirror farther away, you see

      a. more of youself.    c. the same amount of yourself.
      b. less of yourself.

35. Ninety percent of light incident on a certain piece of           (b)BM
    glass passes through it.  How much light passes through two
    pieces of this glass?

      a. 80 %.        d. 89 %.
      b. 81 %.       e. 90 %.
      c. 85 %.

36. Ninety five percent of light incident on a mirror is          (c)BM
    reflected.  How much light is reflected when three of
    these mirrors are arranged so light reflects from one after
    the other?

    a. 81 %.                    d. 90 %.
    b. 85 %.                    e. 95 %.
    c. 86 %.

# 26

## Color

1. Light shines on a pane of green glass and a pane of clear glass. The temperature will be higher in the          (b)A

    a. clear glass.    c. ...neither, it will be the
    b. green glass.       same in each.

2. The colored dots that make up the color on a TV screen are          (b)A

    a. red, blue, yellow.   d. magenta, cyan, yellow.
    b. red, blue, green.    e. red, green, yellow.
    c. yellow, blue, green.

3. Colors seen on TV result from color          (a)A

    a. addition.    c. neither of these.
    b. subtraction.

4. Colors seen on a photograph result from color          (b)A

    a. addition.    c. neither of these.
    b. subtraction.

5. Different colors of light correspond to different light          (d)A

    a. velocities.    d. frequencies.
    b. intensities.    e. none of these.
    c. polarities.

6. In the periphery of our vision, we          (c)A

    a. are more sensitive to low frequecies than high ones.
    b. are insensitive to color and movement.
    c. are sensitive to movement, but cannot see color.
    d. are sensitive to both movement and color.
    e. none of these.

7. The sky is blue because air molecules in the sky act as tiny          (b)A

    a. mirrors which reflect only blue light.
    b. resonators which scatter blue light.
    c. sources of white light.
    d. prisms.
    e. none of these.

8. Stars appear white to us because          (b)A

    a. they are so far away.
    b. their intensity is so low.
    c. the eye cannot see color at night.
    d. they are white.
    e. none of these.

9. Complementary colors are two colors which            (d)A

      a. look good together.
      b. are primary colors.
      c. are next to each other on the color chart.
      d. give white light when added together.
      e. none of these.

10. The complimentary color of blue is                (c)A

      a. red.           d. cyan.
      b. green.         e. magenta.
      c. yellow.

11. Magenta light is really a mixture of             (a)A

      a. red and blue light.   d. yellow and green light.
      b. red and cyan light.   e. none of these.
      c. red and yellow light.

12. The color of an opaque object is determined by light    (c)A
which is

      a. transmitted.      d. all of the above.
      b. absorbed.        e. none of the above.
      c. reflected.

13. Things seen by moonlight usually aren't colored because   (b)B

      a. moonlight doesn't have very many colors in it.
      b. moonlight is too dim to activiate the retina's cones.
      c. moonlight photons don't have enough energy to
         activiate the retina's cones.
      d. all of the above.
      e. none of the above.

14. The main difference between the retina of a human eye and  (b)B
that of a dog's eye is the

      a. predominance of cones in a dog's retina.
      b. absence of cones in a dog's retina.
      c. more intricate optic nerve in a human's eye.
      d. absence of cones in a human's eye.
      e. none of these.

15. A sheet of red paper will look black when illuminated with  (d)B

      a. red light.      d. cyan light.
      b. yellow light.    e. none of these.
      c. magenta light.

16. A blue object will appear black when illuminated with   (c)B

      a. blue light.     d. magenta light.
      b. cyan light.     e. none of these.
      c. yellow light.

17. If sunlight were green instead of white, the most comfortable  (c)B
color to wear on a hot day would be.

      a. magenta.      d. blue.
      b. yellow.       e. none of these.
      c. green.

18. If sunlight were green instead of white, the most comfortable        (a)B
    color to wear on a cold day would be

        a. magenta.              d. blue.
        b. yellow.               e. none of these.
        c. green.

19. The yellow clothes of a stage performer can be made to look          (c)C
    black if illuminated only by light that is

        a. blue.                 c. both of the above.
        b. magenta plus cyan.    d. none of the above.

20. If molecules in the sky scattered orange light instead of            (d)C
    blue light, sunsets would be colored

        a. orange.               d. blue.
        b. yellow.               e. none of these.
        c. green.

21. If the atmosphere were about 50 times thicker, the sun would         (a)C
    appear

        a. red-orange.           d. blue-violet.
        b. orange-green.         e. none of these.
        c. green-blue.

22. The atmosphere of Jupiter is more than 1000 km thick.  From          (d)C
    the planet's surface, the noon-day sun would appear

        a. white.                c. not at all.
        b. faintly white.        d. none of these.

23. Distant dark colored hills appear blue because that is the           (a)C
    color of the

        a. atmosphere between the observer and the hills.
        b. selectively reflected light that reaches a distant
           observer.
        c. selectively refracted light that reaches a distant
           observer.
        d. sky which is reflected off the hills.
        e. none of these.

24. Distant snow covered hills appear yellowish because that             (c)C
    is the color of the

        a. atmosphere between the observer and the hills.
        b. selectively reflected light that reaches a distant
           observer.
        c. selectively refracted light that reaches a distant
           observer.
        d. sky which is relected off the hills.
        e. none of these.

25. On a planet where atmospheric gases are red, distant dark            (d)C
    colored hills would look

        a. bluish.               d. reddish.
        b. greenish.             e. untinted.
        c. yelowish.

26. On a planet where atmospheric gases are yellow, distant              (a)C
    slow covered hills would look

        a. bluish.               d. reddish.
        b. greenish.             e. untinted.
        c. yellowish.

# 27

# Light Waves

1. Diffraction is a result of                                              (c)A

    a. refraction.          d. polarization.
    b. reflection.          e. dispersion.
    c. interference.

2. Newton's rings are a demonstration of                                   (e)A

    a. refraction.          d. polarization.
    b. reflection.          e. interference.
    c. dispersion.

3. Colors seen when gasoline forms a thin film on water are a              (e)A
   demonstration of

    a. refraction.          d. polarization.
    b. reflection.          e. interference.
    c. dispersion.

4. Iridescent colors seen in the pearly luster of an abalone              (e)A
   shell are due to

    a. refraction.          d. polarization.
    b. reflection.          e. none of these.
    c. dispersion.

5. Waves diffract the most when their wavelength is                        (b)A

    a. short.               c. both diffract the same.
    b. long.

6. Waves refract the most when their wavelength is                         (c)B

    a. long.                c. both refract the same.
    b. short.

7. A thin film appears blue when illuminated with white light.            (e)B
   The color being cancelled by destructive interference is

    a. red.                 d. blue.
    b. green.               e. none of these.
    c. white.

8. Light from two very close stars will                                    (b)B

    a. form an interference pattern.
    b. not form an interference pattern.

9. Holograms exist because of                                    (c)B

    a. diffraction.        c. both of the above.
    b. interference.      d. none of the above.


10. Magnification can be accomplished with a hologram if it is    (a)B
    viewed with light that has a

    a. longer wavelength than the original light.
    b. shorter wavelength than the original light.
    c. ...holograms cannot be magnified.


11. Which of the following is a property of light waves, but not  (d)B
    of sound waves?

    a. frequency.        d. polarization.
    b. wavelength.      e. none of these.
    c. amplitude.


12. Interference can be shown using                               (d)B

    a. light waves.      d. all of the above.
    b. sound waves.     e. none of the above.
    c. water waves.


13. The amount of light transmitted from an incandescent lamp     (a)B
    that is transmitted through a real polaroid filter is

    a. less than half.    c. half.
    b. more than half.


14. An ideal polaroid will transmit 50 % of nonpolarized light    (b)C
    incident on it.  How much light is transmitted by two ideal
    polaroids that are oriented with their axes parallel to
    each other?

    a. 0 %.          d. between 0 % and 50 %.
    b. 50 %.        e. between 50 % and 100 %.
    c. 100 %.


15. Because of absorption, a polaroid will actually transmit 40 % (d)C
    of incident nonpolarized light.  Two polaroids with their
    axes aligned will transmit

    a. 0 %.          d. between 0 % and 40 %.
    b. 40 %.        e. between 40 % and 100 %.
    c. 100 %.


16. Because of absorption, a polaroid will actually transmit 40 % (d)CM
    of nonpolarized light incident on it.  Two polaroids with
    their axes aligned will transmit

    a. 16 %.        d. 32 %.
    b. 24 %.       e. 40 %.
    c. 30 %.

# 28

# Light Emission

1. A photographer wishes to use a safety light in the darkroom that will emit low energy photons.  The best color of this light is                                                                                (e)A

   a. violet.          d. yellow.
   b. blue.            e. red.
   c. green.

2. Discrete spectral lines occur when excitation takes place in a          (c)A

   a. solid.           d. superconductor.
   b. liquid.          e. all of these.
   c. gas.

3. Light frequency from an incandescent lamp depends on the                (b)A

   a. amount of electrical energy transformed.
   b. rate of atomic and molecular vibrations.
   c. voltage applied to the lamp.
   d. electrical resistance of the lamp.
   e. transparency of glass.

4. Which color light carries the most energy?                              (e)A

   a. red.             d. green.
   b. orange.          e. blue.
   c. yellow.

5. Ultraviolet light is                                                    (c)A

   a. more energetic than X-rays.
   b. produced by crossed polaroids.
   c. electromagnetic energy.
   d. present everywhere.
   e. none of these.

6. Compared to the energy of a photon of red light, the energy of a photon of blue light is                                                (b)A

   a. less.            c. the same.
   b. more.

7. Isolated bells ring clear, while bells crammed in a box ring muffled.  If the sound of isolated bells is analogous to light from a gas discharge tube, then sound from the box crammed with bells is analogous to light from                    (c)A

   a. a laser.              d. a phosphorescent source.
   b. a fluroescent lamp.   e. none of these.
   c. an incandescent lamp.

8. Compared to the energy put into a laser, the energy of the laser beam is                                                      (b)A

    a. more.             c. the same.
    b. less.

9. The radiation curve for a "red hot" object peaks in the                                                      (a)B

    a. infrared region.    d. ultraviolet region.
    c. green region.

10. The radiation curve for a "blue hot" object peaks in the                                                      (d)B

    a. infrared region.    d. ultraviolet region.
    b. red region.        e. none of these.
    c. yellow region.

11. A paint pigment that absorbs red light and gives off blue light                                                      (e)B

    a. is fluorescent.     d. is polarized.
    b. is phosphorescent.  e. can't exist.
    c. is used in lasers.

12. The red laser beam from a helium neon laser corresponds to a spectral line of                                                      (b)B

    a. helium.        c. both of the above.
    b. neon.         d. none of the above.

13. A condition necessary for absorption spectra to exist is that                                                      (b)B

    a. the light source be a gas.
    b. partially absorbent material exist between the light
       source and spectroscope.
    c. the spectroscope be equipped with an absorption cell.
    d. all of the above.
    e. none of the above.

14. Which light source is a more energy efficient?                                                      (a)B

    a. a fluorescent lamp.  c. both are the same.
    b. an incandescent lamp.

15. If light in a spectroscope passed through round holes instead of slits, spectral lines would appear                                                      (b)B

    a. as lines.       c. square.
    b. round.

16. The energy of a photon depends on its                                                      (b)B

    a. speed.         d. amplitude.
    b. frequency.     e. none of these.
    c. E-field orientation.

17. Some minerals glow when illuminated with ultraviolet light. This is because                                                      (a)B

    a. ultraviolet photons kick atomic electrons in the
       mineral into higher energy states.
    b. ultraviolet photons have such high energy.
    c. of selective reflection.
    d. of selective transmission.
    e. none of these.

18. Some light switches glow in the dark after the lights are         (c)B
    turned off.  This is because of

        a. fluorescence.        d. incandescence.
        b. resonance.           e. none of these.
        c. a time delay between excitation and de-excitation.


19. A laser emits light because of                                    (b)B

        a. scattering.          d. incandescence.
        b. de-excitation.       e. none of these.
        c. interference.


20. Materials can be heated until "red hot".  If some material        (c)C
    is heated until it is "green hot", then

        a. it would liquify immediately.
        b. it would be hotter than "white hot".
        c. its molecules would be vibrating at identical rates.
        d. the red part of the radiation would be eliminated.
        e. none of these.


21. If the radiation curve for an object peaks in the green           (e)C
    region, the object would appear

        a. red.                 d. blue.
        b. yellow.              e. white.
        c. green.

# 29

# Light Quanta

1. Which has less energy per photon?                                    (a)A

    a. red light.        c. both have the same energy.
    b. blue light.

2. Which has more energy per photon?                                    (b)A

    a. red light.        c. both have the same energy.
    b. blue light.

3. Which of the following photons have the greatest energy?             (e)A

    a. infrared.        d. blue light.
    b. red light.       e. ultraviolet.
    c. green light.

4. The photoelectric effect best demonstrates the                      (b)A

    a. wave nature of light.
    b. particle nature of light.
    c. both of the above.
    d. none of the above.

5. In the photoelectric effect, the brighter the illuminating          (a)A
   light on a photosensitive surface, the greater the

    a. number of ejected electrons.
    b. velocity of ejected electrons.
    c. both of the above.
    d. none of the above.

6. A lump of energy associated with light is called a                  (c)A

    a. quantum.       c. both of the above.
    b. photon.        d. none of the above.

7. A photocell can be activated with red light, but not with           (b)A
   blue light.  This statement is

    a. true.         b. false.

8. The ratio of a photon's energy to its frequency is                  (d)A

    a. its speed.       d. Planck's constant.
    b. its wavelength.    e. none of these.
    c. its amplitude.

9. In the relationship E = hf for a photon emitted from an atom, the symbol E is used to represent the energy      (c)B

    a. of the emitted photon.
    b. difference between atomic energy states producing
       the photon.
    c. both of the above.
    d. none of the above.


10. A photosensitive surface is illiminated with both blue and violet light. The light that will cause the most electrons to be ejected is the      (d)B

    a. blue light.          c. both eject the same number.
    b. violet light.        d. not enough information given.


11. Which of the following is not quantized?      (e)B

    a. energy.              d. electric charge.
    b. radiation.           e. ...all are quantized.
    c. number of people in a room.


12. Two photons have the same wavelength. They also have the same      (c)B

    a. frequency.           c. both of the above.
    b. energy.              d. none of the above.


13. When a clean surface of potassium metal is exposed to blue light, electrons are emitted. If the intensity of the blue light is increased, which of the following will also increase?      (a)B

    a. the number of electrons ejected per second.
    b. the maximum kinetic energy of the ejected electrons.
    c. the threshold frequency of the ejected electrons.
    d. the time lag between the absorption of blue light
       and the start of emission of the electrons.
    e. none of these.


14. In the photoelectric effect, electrons ejected from bound states in the photosensitive material have      (a)C

    a. less kinetic energy than the absorbed photon's energy.
    b. more kinetic energy than the absorbed photon's energy.
    c. kinetic energy equal to the absorbed photon's energy.

# 30

# The Atom
# and the Quantum

Chapter 30: Atoms and Quanta

1. Quantization of electron energy states in an atom is better     (a)A
understood in terms of the electron's

    a. wave nature.          c. neither of these.
    b. particle nature.

2. An excited hydrogen atom is capable of emitting radiation of     (c)A

    a. 3 frequencies.        c. more than 3 frequencies.
    b. a single frequency.

3. The Schroedinger equation provides an adequate model for     (d)A

    a. submicoscopic particles.
    b. microscopic particles.
    c. macroscopic particles.
    d. all of the above.
    e. none of the above.

4. In the Bohr model of hydrogen, discrete radii and energy     (d)A
states result when an electron circles the atom in
an integral number of

    a. wave frequencies.     d. de Broglie wavelengths.
    b. quantum numbers.      e. none of these.
    c. diffraction patterns.

5. Which of the following forms an interference pattern when     (d)A
directed towards two suitably-spaced slits?

    a. light.                d. all of the above.
    b. sound.                e. none of the above.
    c. electrons.

6. A beam of electrons has     (c)A

    a. wave properties.      c. both of the above.
    b. particle properties.  d. none of the above.

7. Compared to the average diameter of a hydrogen atom, the     (b)A
average diameter of a helium atom is

    a. larger.               c. about the same.
    b. smaller.

8. According to the correspondence principle, a new theory is valid if it    (d)A

    a. be valid where the old theory works.
    b. accounts for results from the old theory.
    c. gives the same correct results as the old theory.
    d. all of the above.
    e. none of the above.

9. The correspondence principle applies to    (c)A

    a. submicroscopic theories.
    b. macroscopic theories.
    c. all good theories.

10. Compared to the diameter of a zirconium atom (A = 40), the diameter of a mercury atom (A = 80) is approximately    (c)A

    a. four times as large.    d. one half as large.
    b. twice as large.    e. one quarter as large.
    c. the same size.

11. Planck's constant is a basic constant of nature that    (b)A

    a. sets an upper limit on the size of things.
    b. sets a lower limit on the size of things.
    c. formulates the relationship between mass and energy.
    d. relates the energy of a photon to its momentum.
    e. is the foundation of the correspondence principle.

12. Orbital electrons do not spiral into the nucleus because of    (d)A

    a. electromagnetic forces.
    b. angular momentum conservation.
    c. the large nuclear size compared to the electron's size.
    d. the wave nature of the electron.
    e. none of these.

13. The main reason electrons occupy discrete orbits in an atom is because    (d)A

    a. energy levels are quantized.
    b. electric forces act over quantized distances.
    c. electrons are basically discrete particles.
    d. the circumference of each orbit must be an integral multiple of electron wavelengths.
    e. none of these.

14. A new theory conforms to the correspondence principle when it    (d)A

    a. corresponds to all theories in nature.
    b. updates the essence of the old theory.
    c. ties two or more theories together.
    d. accounts for verified results of the old theory.
    e. none of these.

15. The quantity most common to a proton and an electron is    (b)A

    a. mass.    d. all of the above.
    b. charge.    e. none of the above.
    c. energy.

16. The ratio of the energy of a photon to its frequency is    (b)A

    a. pi.    d. the photon's wavelength.
    b. Planck's constant.    e. not known.
    c. the photon's speed.

17. The electrical force between an inner electron and the        (b)A
    nucleus of an atom is larger for atoms of

        a. low atomic number.    c. ...the same for both.
        b. high atomic number.

18. According to the uncertainty principle, the more we know      (d)B
    about a particle's momentum, the less we know about its

        a. kinetic energy.    d. location.
        b. mass.              e. none of these.
        c. speed.

19. According to the uncertainty principle, the more we know      (d)B
    about a particle's position, the less we know about its

        a. speed.             d. all of the above.
        b. momentum.          e. none of the above.
        c. kinetic energy.

20. The wavelengths of matter waves are relatively               (b)B

        a. large.             b. small.

21. An excited atom decays to its ground state and emits a photon (a)B
    of green light.  If instead the atom decays to an intermed-
    iate state, then the light emitted could be

        a. red.               d. any of the above.
        b. violet.            e. none of the above.
        c. blue.

22. An electron and a baseball move at the same speed.  Which has (a)C
    the longer wavelength?

        a. the electron.      c. both have the same
        b. the baseball.         wavelength.

23. A bullet and a proton have the same momentum.  Which has the  (b)C
    longer wavelength?

        a. the bullet.        c. both have the same
        b. the proton.           wavelength.

24. The wavelength of a moving basball is very                    (a)C

        a. short.             c. non-existent.
        b. long.

25. A hypothetical atom has four distinct energy states.  Assuming (b)C
    all transitions are possible, how many spectral lines can
    this atom produce?

        a. 5.                 d. 8.
        b. 6.                 e. more than 8.
        c. 7.

26. An electron and a proton are traveling at the same speed.     (a)C
    Which has a longer wavelength?

        a. the electron.      c. both have the same
        b. the proton.           wavelength.

27. Which of the following has the longer wavelength?                    (a)C

    a. a low energy electron.
    b. a high energy electron.
    c. both have the same wavelength.

# 31

# The Atomic Nucleus and Radioactivity

1. X-rays may be regarded as                                          (b)A

    a. high frequency sound waves.
    b. high frequency radio waves.
    c. both of the above.
    d. none of the above.

2. X-rays are similar to                                              (c)A

    a. alpha rays.       d. all of the above.
    b. beta rays.        e. none of the above.
    c. gamma rays.

3. Which radiation has no electric charge associated with it?         (c)A

    a. alpha rays.       d. all of the above.
    b. beta rays.        e. none of the above.
    c. gamma rays.

4. An atom with an imbalance of electrons to protons is               (c)A

    a. a hadron.        d. an isotope.
    b. a baryon.        e. none of these.
    c. an ion.

5. Electric forces within an atomic nucleus tend to                   (b)A

    a. hold it together.   c. neither of these.
    b. push it apart.

6. Generally speaking, the larger a nucleus is, the more it is        (b)A

    a. stable.        c. neither stable nor
    b. unstable.        unstable.

7. The halflife of an isotope is one day.  At the end of two          (c)A
days how much of the isotope remains?

    a. none.         d. one eighth.
    b. one half.       e. none of these.
    c. one quarter.

8. The halflife on an isotope is one day.  At the end of              (d)A
three days, how much of the isotope remains?

    a. none.         d. one eighth.
    b. one half.       e. none of these.
    c. one quarter.

9. The half-life of a radioactive substance is independent of (e)A

    a. the number (if large enough) of atoms in the
       substance.
    b. whether the substance exists in an elementary state
       or in a compound.
    c. the temperature of the substance.
    d. the age of the substance.
    e. all of these.

10. Operation of a cloud chamber relies on (e)A

    a. magnetization.      d. polarization.
    b. evaporation.        e. condensation.
    c. acceleration.

11. When a nucleus emits a beta particle, its atomic number (c)A

    a. remains constant, but its mass number changes.
    b. remains constant, and so does its mass number.
    c. changes, but its mass number remains constant.
    d. changes, and so does its mass number.
    e. none of these.

12. Carbon 14 is produced in the atmosphere principally by (b)A

    a. plants and animals.
    b. cosmic ray bombardment.
    c. nitrogen bombardment.
    d. photosynthesis.
    e. none of these.

13. Which experiences the greatest electrical force in an (a)A
    electric field?

    a. alpha particle.      d. gamma ray.
    b. beta particle.       e. all deflected the
    c. electron.               same.

14. Which experiences the least electrical force in an electric (d)A
    field?

    a. alpha particle.      d. gamma ray.
    b. beta particle.       e. all affected the
    c. electron.               same.

15. Carbon dating requires that the object being tested contains (a)A

    a. organic material.    d. sugar molecules.
    b. inorganic material.  e. none of these.
    c. charcol.

16. Radiation damage to living tissue is detrimental (b)A

    a. only when cells are killed.
    b. when cells are damaged as well as killed.
    c. ...radiation is not detrimental to living tissue.

17. Most radiation in the air comes from (d)A

    a. nuclear power plants.
    b. weapons testing fallout.
    c. medical X-rays.
    d. cosmic rays and Earth minerals.
    e. none of these.

18. Where do cosmic rays originate?                              (d)A

    a. the Earth.         d. the cosmos.
    b. clouds.           e. none of these.
    c. the sun.

19. When radium (A = 88) emits an alpha particle, the resulting    (a)B
nucleus has atomic number

    a. 86.           d. 92.
    b. 88.           e. none of these.
    c. 90.

20. When thorium (A = 90 ) emits a beta particle, the resulting    (e)B
nucleus has atomic number

    a. 88.           d. 92.
    b. 89.           e. none of these.
    c. 90.

21. When a nucleus emits a positron, its atomic number            (b)B

    a. increases by 1.    c. doesn't change.
    b. decreases by 1.

22. When a nucleus emits a beta particle, its atomic number        (a)B

    a. increases by 1.    d. decreases by 2.
    b. decreases by 1.    e. none of these.
    c. increases by 2.

23. A gram of radioactive material has a half life of one year.    (c)B
After 3 years, how much radioactive material will be left?

    a. 0 g.          d. 1/4 g.
    b. 1/16 g.      e. none of these.
    c. 1/8 g.

24. Artificially induced radioactive elements generally have       (b)B

    a. long half lives.    c. medium length half
    b. short half lives.       lives.

25. A sample of radioactive material is somewhat                   (c)B

    a. cooler than its surroundings.
    b. the same temperature as its surroundings.
    c. warmer than its surroundings.

26. It's impossible for a hydrogen atom to emit an alpha particle.  (a)B

    a. true.          b. false.

27. If an alpha particle and a beta particle have the same energy,  (b)B
which particle will penetrate the farthest in an object?

    a. alpha particle.    c. they both penetrate
    b. beta particle.      the same distance.

28. An element will decay to an element with higher atomic          (a)B
number in the periodic table if it emits

    a. a beta particle.    d. an alpha particle.
    b. a gamma ray.      e. none of these.
    c. a proton.

29. The end result of radioactive decay can be a different          (d)B

        a. isotope.              d. all of the above.
        b. element.             e. none of the above.
        c. atom.

30. The half life of carbon 14 is 5730 years. If a 1 gram          (c)BM
    sample of old carbon is 1/8 as radioactive as 1 gram of a
    current sample, then the age of the old sample is

        a. 716 years.           d. 22920 years.
        b. 11460 years.         e. none of these.
        c. 17190 years.

31. A certain radioactive isotope placed near a geiger counter     (d)BM
    registers 160 counts per second. Eight hours later, the
    counter registers 10 counts per second. What is the half
    life of the isotope?

        a. 8 hours.             d. 2 hours.
        b. 6 hours.             e. none of these.
        c. 4 hours.

32. A certain radioactive isotope placed near a geiger counter     (d)BM
    registers 120 counts per minute. If the half life of the
    isotope is one day, what will the count rate be at the end
    of four days?

        a. 30 counts/min.       d. 7.5 counts/min.
        b. 15 counts/min.       e. 5 counts/min.
        c. 10 counts/min.

33. A geiger counter placed 1 meter from a point source of         (e)BM
    radiation registers 100 counts per second. If the geiger
    counter is moved closer to 0.5 meter from the source, what
    will the count rate be?

        a. 25 counts/min.       d. 200 counts/min.
        b. 50 counts/min.       e. 400 counts/min.
        c. 100 counts/min.

34. In bubble chambers, charged particles move in spirals because  (e)C

        a. the magnetic field decreases.
        b. the electric charge decreases.
        c. the electric charge increases.
        d. of perspective and parallax.
        e. of energy dissipation.

35. In order for an atom to decay to an element which is one       (d)C
    greater in atomic number, it can emit

        a. one alpha particle and 3 beta particles.
        b. one positron and 2 beta particles.
        c. one beta particle.
        d. all of the above.
        e. none of the above.

36. An element emits 1 alpha particle, 1 positron, and 3 beta      (c)C
    particles. Its atomic number

        a. decreases by 2.      d. increases by 1.
        b. decreases by 1.      e. increases by 2.
        c. stays the same.

# 32

# Nuclear Fission and Fusion

1. Electrical forces inside a nucleus contribute nuclear          (b)A

    a. stability.       c. both of the above.
    b. unstability.     d. none of the above.

2. Uranium 235, uranium 238 and uranium 239 are different          (c)A

    a. elements.       c. isotopes.
    b. ions.         d. none of these.

3. Different isotopes of an element have different numbers of          (d)A

    a. protons.      d. neutrons.
    b. hadrons.      e. none of these.
    c. photons.

4. In nuclear fission and nuclear fussion reactions, the amount          (a)A
of mass converted to energy is about

    a. less than 1 %.    d. 30 %.
    b. 10 %.        e. more than 30%.
    c. 20 %.

5. Present day research in nuclear fusion uses          (d)A

    a. laser beams.     d. all of the above.
    b. ion beams.      e. none of the above.
    c. ultra hot plasmas.

6. When two light atoms fuse together, mass          (a)A

    a. is converted to energy.
    b. is created from energy.
    c. remains the same.
    b. is gained.

7. A nuclear process which has relatively few radioactive          (b)A
by-products is

    a. fission.       c. both of these have lots of
    b. fusion.          radioactive by-products.

8. In both fission and fusion, mass          (b)A

    a. is created from energy.
    b. is changed into energy.
    c. remains the same.

9. Detonation of a fission type atomic bomb is started by (b)A

    a. splitting a small piece of uranium.
    b. pressing together several small pieces of uranium.
    c. igniting a small thermonuclear bomb.
    d. turning on a laser cross fire.
    e. none of these.

10. In a breeder reactor (b)A

    a. Plutonium is converted into Uranium.
    b. Uranium is converted into Plutonium.
    c. both of the above.
    d. none of the above.

11. Radioactive by-products are the result of atomic nuclei (b)A

    a. breaking apart.      c. both of the above.
    b. combining.           d. none of the above.

12. Energy released by the sun results from atomic nuclei (b)A

    a. breaking apart.      c. both of the above.
    b. combining.           d. none of the above.

13. Between nuclear fission and nuclear fusion, radioactive (a)A
    by-products are more characteristic of nuclear

    a. fission.             c. both of the above.
    b. fusion.              d. none of the above.

14. If all the Uranium in the world were exhausted, breeder (d)A
    reactors would be

    a. used much more than they are now.
    b. researched more than they are now.
    c. considered more seriously than they are now.
    d. relics of a brief age.
    e. none of these.

15. Hydrogen is a first-rate fuel for (c)A

    a. nuclear reactions.   c. both of the above.
    b. chemical reactions.  d. none of the above.

16. The mass of a nucleus is exactly equal to the sum of the (e)B
    masses of its individual

    a. protons.             d. all of the above.
    b. neutrons.            e. none of the above.
    c. nucleons.

17. In gaseous form, at the same temperature, the average speed of (a)B
    U-238 compared to the speed of U-235 is

    a. less.                c. the same.
    b. more.

18. Reactions that takes place in a breeder reactor change (a)B

    a. elements into different elements.
    b. molecules into different molecules.
    c. ions into different ions.
    d. all of the above.
    e. none of the above.

19. A nuclear proton has greatest mass in                          (c)B

        a. iron.                 d. uranium.
        b. oxygen.               e. same in each
        c. hydrogen.                of these.

20. A nuclear proton has least mass in                             (a)B

        a. iron.                 d. uranium.
        b. oxygen.               e. same in each
        c. hydrogen.                of these.

21. The mass of a nucleon is greatest when located                 (b)B

        a. inside a nucleus.    c. same in both cases.
        b. outside a nucleus.

22. Fissioning helium would yield a net                            (a)B

        a. absorption of energy.
        b. release of energy.
        c. neither absorption nor release of energy.

23. Fusing two helium nuclei yields a net                          (b)B

        a. absorption of energy.
        b. release of energy.
        c. neither absorption nor release of energy.

24. Fissioning an iron nucleus yields a net                        (a)B

        a. absorption of energy.
        b. release of energy.
        c. neither absorption nor release of energy.

25. Fusing two iron nuclei yields a net                            (a)B

        a. absorption of energy.
        b. release of energy.
        c. neither absorption nor release of energy.

26. Fissioning a lead nucleus yields a net                         (b)B

        a. absorption of energy.
        b. release of energy.
        c. neither absorption nor release of energy.

27. The reason neutrons make better nuclear bullets in nuclear     (e)B
    reactions is because a neutron has

        a. less mass than a proton.
        b. more penetrating power than a proton.
        c. a larger cross section than a proton.
        d. all of the above.
        e. none of the above.

28. The reason plutonium is not found in natural ore deposits is   (d)B
    because it

        a. is artificially created.
        b. is chemically inert.
        c. is a gas at room temperature.
        d. has a short half life.
        e. none of these.

29. Compared to the sum of the masses of all the individual     (b)B
    nucleons that make up a nucleus, the mass of the nucleus is

        a. more.                c. the same.
        b. less.

30. If gold were used as nuclear fuel, it would be best         (b)B

        a. fused.               c. either fused or
        b. fissioned.              fissioned.

31. If carbon were used as nuclear fuel, it would be best       (a)B

        a. fused.               c. either fused or
        b. fissioned.              fissioned.

32. If iron were used as nuclear fuel, it would be best         (c)B

        a. fused.               c. neither fused nor
        b. fissioned.              fissioned.

33. Compared to the energy produced by fissioning a single uran-  (b)B
    ium atom, the energy produced by fusing two deuterium atoms is

        a. more.                c. the same.
        b. less.

34. Compared to the energy produced by fissioning a gram of uran-  (a)B
    ium, the energy produced by fusing a gram of deuterium is

        a. more.                c. the same.
        b. less.

35. In which of these processes is an element of matter changed    (c)B
    into a completely different element?

        a. nuclear fission.     c. both of the above
        b. nuclear fusion.      d. none of the above.

36. Suppose a hydrogen bomb were exploded in a box that could      (c)C
    contain all the energy released by the explosion.  Compared
    to the weight of the box before the explosion, the weight
    of the box after the explosion would be

        a. more.                c. the same.
        b. less.

37. Which shape uses the smallest amount of material when          (d)C
    creating a critical mass?

        a. cube.                d. sphere.
        b. elongated box.       e. none of these.
        c. cone.

38. Which element has the smaller critical mass – uranium 235      (b)C
    which releases 2.5 neutrons per fission, or plutonium
    which releases 2.7 neutrons per fission?

        a. uranium 235.         c. both would have the
        b. plutonium.              same critical mass.

39. Compared to a neutron bouncing off a gold nucleus, a           (b)C
    neutron bouncing off a hydrogen nucleus loses

        a. less speed.          c. the same amount
        b. more speed.             of speed.

# 33

# The Special Theory of Relativity

1. According to Einstein's theory of special relativity,                    (c)A

    a. space and time are aspects of each other.
    b. energy and mass are aspects of each other.
    c. both of the above.
    d. none of the above.

2. According to the special theory of relativity, all laws of              (c)A
nature are the same in reference frames that

    a. accelerate.        d. oscillate.
    b. move in circles.    e. none of these.
    c. move at constant speeds.

3. Objects accelerated to relativistic speeds appear to                    (c)A

    a. live shorter.     d. all of the above.
    b. grow bigger.      e. none of the above.
    c. gain mass.

4. Compared to clocks in a stationary reference frame, clocks              (a)A
in a moving reference frame run

    a. slower.          c. at the same speed.
    b. faster.

5. A woman standing on the ground sees a rocket ship move past             (b)A
her at 95 % the speed of light.  Compared to when the
rocket is at rest, the woman sees the rocket's length as

    a. longer.         c. the same length.
    b. shorter.

6. Clocks on a space ship moving very fast appear to run slow              (b)A
when viewed from

    a. the space ship.    c. both places.
    b. the Earth.       d. neither places.

7. When an object is pushed to relativistic speeds, its                    (a)A
mass is measured to be

    a. greater than at rest.
    b. smaller than at rest.
    c. the same as at rest.

8. According to the well known equation, energy equals mass       (d)A
   times the speed of light squared,

       a. mass and energy travel at the speed of light
          squared.
       b. energy is actually mass traveling at the speed of
          light squared.
       c. mass and energy travel at twice the speed of light.
       d. mass and energy are related.
       e. none of these.


9. Relativity equations for time, length and mass hold true for   (c)A

       a. relativistic speeds.
       b. everyday low speeds.
       c. both of the above.
       d. none of the above.


10. When you approach a light source, its speed appears to        (c)A

       a. increase.          c. stay the same.
       b. decrease.


11. When you approach a light source which in turn is moving      (c)A
    towards you, the relative speed between you and the emitted
    light waves

       a. increases.          c. stays the same.
       b. decreases.


12. According to the special theory of relativity, if you         (c)B
    measure your pulse while traveling at very high speeds, you
    would notice your pulse rate to

       a. increase.          c. be the same as
       b. decrease.             usual.


13. If you were to travel at speeds close to the speed of light,  (e)B
    you would notice that your

       a. mass increases.     d. all of the above.
       b. pulse decreases.    e. none of the above.
       c. shape changes.


14. Densities measured by observers in stationary reference       (a)B
    frames of objects that are moving at relativistic speeds
    are

       a. larger than when measured at rest.
       b. smaller than when measured at rest.
       c. the same as when measured at rest.


15. A spaceship that is traveling very fast with respect to       (c)B
    your frame of reference, fires a photon beam than travels
    at  speed c with respect to the spaceship.  You measure the
    photon beam's speed to be

       a. less than c.        c. equal to c.
       b. more than c.


16. We are actually looking into the past when we look at         (c)B

       a. starlight.          c. actually both of the above.
       b. our physics book.   d. none of the above.

17. When you approach a light source, the wavelength of light    (b)B
    emitted appears

        a. longer.              c. the same.
        b. shorter.

18. The frequency of a light source doubles as the light         (a)B
    approaches you. As the same light source moves away from
    you at the same speed, its frequency

        a. is halved.           c. stays the same.
        b. is doubled.

19. As a blinking light source approaches you at an increasing   (a)B
    speed, the frequency of the flashes

        a. increases.           c. stays the same.
        b. decreases.

20. In some reference frame in the universe, you, right now,     (a)B
    are traveling at speeds close to the speed of light.

        a. true.                b. false.

21. To outside observers, objects traveling at relativistic      (b)B
    speeds look

        a. larger.              c. the same size.
        b. smaller.

22. A spaceship whizzes past a space station at a very high      (b)B
    speed relative to the station. An observer on the space
    station measures the mass of the spaceship larger than when
    the ship is at rest and an observer on the spaceship
    measures the mass of the space station to be

        a. less than if the spaceship weren't moving.
        b. more than if the spaceship weren't moving.
        c. the same as when the observer isn't moving.

23. Suppose you and your sister travel in space in such a way    (b)B
    that you notice your sister's time runs slower than yours.
    Your sister will notice that your time runs

        a. faster than hers.    c. the same as hers.
        b. slower than hers.    d. not enough information given.

24. Compared to Earth time, there is a physical slowing of time  (c)B
    when you travel at

        a. relativistic speeds.
        b. everyday low speeds.
        c. both of the above.
        d. none of the above.

25. A heavy meter stick has a rest mass of 1 kg. When the        (d)C
    meter stick is thrown past you, you measure its mass to be
    2 kg. Assuming the meter stick is thrown horizontally, what
    do you measure its length to be?

        a. 4 m.                 d. 0.5 m.
        b. 2 m.                 e. none of these.
        c. 1 m.

26. There is an upper limit on the speed of a particle. This          (d)C
    means there is also an upper limit on its

        a. momentum.              c. both of the above.
        b. kinetic energy.        d. none of the above.


27. Suppose a blinking light source uniformly accelerates away        (b)C
    from you. As the source gets farther and farther away,
    you'll notice that the frequency of the flashes

        a. increases.             c. neither increases nor
        b. decreases.                decreases.


28. A 10 meter long spear is thrown at relativistic speeds            (d)C
    through a 10 meter long pipe. (Both these dimensions
    are measured when each is at rest.) When the spear passes
    through the pipe, which of the following statements is true?

        a. the spear shrinks so the pipe completely covers it.
        b. the pipe shrinks so the spear extends from both ends.
        c. both shrink equally so the pipe barely covers the
           spear.
        d. any of the above, depending on the motion of the observer
           (moving with the spear, at rest with the pipe, etc.).
        e. none of the above.

# 34

# The General Theory of Relativity

1. Compared to special relativity, general relativity is more        (d)A
   concerned with

   a. acceleration.          d. all of the above.
   b. gravitation.           e. none of the above.
   c. space time geometry.

2. According to the principle of equivalence, observations made      (e)A
   in a Newtonian gravitational field are indistinguishable from
   observations made in

   a. any other gravitational field.
   b. an Einsteinian gravitational field.
   c. any uniformly moving reference frame.
   d. all of the above.
   e. none of the above.

3. From a general relativistic point of view, a person on the       (b)A
   ground floor of a skyscraper ages

   a. faster than a person on the top floor.
   b. slower than a person on the top floor.
   c. at the same speed as a person on the top floor.

4. A measuring stick on a rapidly rotating disk will not            (b)A
   appear to shrink if it is oriented along the

   a. circumference.         c. either of the above.
   b. radius.                d. none of the above.

5. The measured ratio of circumference to diameter on a rapidly     (c)A
   rotating disk is pi when the disk is

   a. moving at relativistic speeds.
   b. moving at any speed.
   c. at rest.
   d. all of the above.
   e. none of the above.

6. The two-dimensional surface of the Earth is                      (a)A

   a. positively curved.     c. both of the above.
   b. negatively curved.     d. none of the above.

7. Compared to Newton's theory of gravitation, Einstein's           (b)A
   theory

   a. is an exception to the correspondence principle.
   b. obeys the correspondence principle.
   c. neither of these.

8. In a one-g gravitational field, in one second, a light beam          (b)A
   will curve beneath a perfectly straight line by

        a. less than 4.9 m.      c. more than 4.9 m.
        b. 4.9 m.


9. At the top of very tall skyscraper, a sensitive watch                (b)A
   will appear to run

        a. slow.                 c. the same as at 1 g.
        b. fast.


10. Compared to a watch at the Earth's poles, a watch at the            (b)A
    Earth's equator should run

        a. a tiny bit slower.    c. at the same speed.
        b. a tiny bit faster.


11. If a star that is 10 lightyears away from Earth explodes,           (c)B
    gravitational waves from the explosion will reach the Earth
    in

        a. less than 10 years.   c. 10 years.
        b. more than 10 years.


12. At the bottom of a well deep enough so that the                     (b)B
    acceleration due to gravity is less than 1 g, a watch will
    appear to run

        a. slow.                 c. the same as at 1 g.
        b. fast.


13. If a star 100 light years away explodes, time for the              (e)B
    gravitational waves to reach the Earth is 100 years as
    measured from

        a. any uniformly moving reference frame.
        b. any accelerating reference frame.
        c. any reference frame, accelerated or not.
        d. the reference frame of the waves.
        e. the Earth.


14. A clock on the surface of a shrinking star will run                 (a)B
    progressively

        a. slower.               c. ...no difference.
        b. faster.


15. Light bends when it                                                 (c)B

        a. passes a massive star.
        b. passes through a gravitational field.
        c. both of the above.
        d. none of the above.


16. The quantity that undergoes a red shift is                          (c)B

        a. wave frequency.       c. both of the above.
        b. wavelength.           d. none of the above.


17. The quantity that shifts in a gravitational red shift is            (a)B

        a. wave frequency.       d. spacetime curvature.
        b. wave direction.       e. none of these.
        c. field intensity.

18. The orbit of Mercury precesses because (d)B

    a. Mercury moves in the gravitational field of the other planets.
    b. Mercury travels faster than any other planet.
    c. Mercury is closest to the Sun.
    d. the Sun's gravitational field varies along Mercury's orbit.
    e. none of these.

19. Einstein's theory of gravitation obeys the correspondence (b)B
principle because it

    a. corresponds to truer description of events in very large gravitational fields.
    b. agrees with proven results of Newton's theory.
    c. has been proven in repeated experiments.
    d. is a special case of Newton's theory.
    e. none of these.

20. According to the principle of equivalence, (d)B

    a. mass and energy are two forms of the same thing.
    b. space and time are two forms of the same thing.
    c. electricity and magnetism are two forms of the same thing.
    d. observations made in an accelerating reference frame are indistinguishable from those made in a gravitational field.
    e. all of these.

21. According to relativity theory, it is possible to grow (d)B
younger when you

    a. are near a block hole.
    b. are near a very large gravitational field.
    c. travel at nearly the speed of light.
    d. ...you can never grow younger.
    e. none of these.

22. If the orbit of Mercury were perfectly circular, its rate of (d)C
precession would be

    a. larger.        c. the same.
    b. smaller.       d. zero.

23. If the elliptical orbit of Mercury were more eccentric, its (a)C
precession rate would be

    a. larger.        c. the same.
    b. smaller.       d. nonexistent.

24. From a relativistic point of view, light (b)C

    a. always travels in straight lines.
    b. sometimes travels in straight lines.
    c. never travels in straight lines.

25. An astronaut falling into a black hole would see the (b)C
universe

    a. red shifted.      b. blue shifted.

26. If the Sun collapsed to a black hole, the time required for (c)C
the Earth to orbit the collapsed Sun would

    a. increase.      c. stay the same.
    b. decrease.

# 35

# Astrophysics

Chapter 35: Astrophysics

1. Right now the universe is expanding at a rate that is  (b)A

    a. speeding up.          c. constant.
    b. slowing down.

2. Stars do not ordinarily collapse because of  (c)A

    a. gravitational shielding.
    b. inverse gravitational forces.
    c. radiation pressure.
    d. all of the above.
    e. none of the above.

3. Stars that are more short-lived usually are stars that are  (a)A

    a. more massive than other stars.
    b. less massive than other stars.
    c. ...a star's life doesn't depend on mass.

4. Compared to younger stars, the Sun spins  (b)A

    a. faster.               c. at the same rate.
    b. slower.

5. Most of the solar system's angular momentum is found in the  (d)A

    a. Sun.                  d. planets.
    b. Asteroids.            e. none of these.
    c. comets and cosmic rays.

6. Most of the mass in the solar sytem is found in the  (a)A

    a. Sun.                  d. planets.
    b. Asteroids.            e. none of these.
    c. comets and cosmic rays.

7. Stars having the greatest angular momentum are usually  (a)A

    a. young stars.          d. novas.
    b. middle aged stars.    e. none of these.
    c. old stars.

8. Elements heavier than iron that we find on Earth,  (b)A
   originated in

    a. the Sun.              d. neighboring galaxies.
    b. supernovae.           e. none of these.
    c. ancient star cores.

9. After a supernova explosion, what's left at the center of      (d)A
   the expanding gases is a

    a. white dwarf.       d. neutron star.
    b. black dwarf.      e. none of these.
    c. red dwarf.

10. An abundance of heavy elements is found in           (b)A

    a. old stars.        d. all of the above.
    b. young stars.     e. none of the above.
    c. cosmic gas.

11. When a star collapses into a black hole, its mass    (c)A

    a. increases.        c. stays about the
    b. decreases.           same.

12. The dominant force in the universe is the         (c)A

    a. nuclear force.    d. electromagnetic force.
    b. weak force.      e. none of these.
    c. gravitational force.

13. A photon sphere is                          (d)A

    a. the outer envelope of a star's atmosphere.
    b. a special case of a photon ellipsoid.
    c. a globular cluster of trapped photons.
    d. photons in circular orbit about a black hole.
    e. none of these.

14. The brightest objects known in the universe today are  (e)A

    a. pulsars.         d. supernovae.
    b. black holes.     e. quasars.
    c. white dwarfs.

15. The star that is more likely to have a system of planets is  (b)A
   one that is rotating

    a. fast.          b. slow.

16. A protostar most easily collpases to form a star if its  (a)B
   mass is

    a. large.        c. mass is not involved in
    b. small.          protostar collapse.

17. In general, the brightness of a star becomes       (a)B

    a. increases with age.  c. neither increases nor
    b. decreases with age.    decreases with age.

18. The Sun will not produce carbon in its core because it  (c)B
   lacks enough

    a. momentum.      d. time.
    b. heat.         e. none of these.
    c. mass.

19. If nuclear fusion in the Sun stopped, the Sun would    (b)B

    a. explode.       c. stay as it is.
    b. implode.

20. The star most likely to have a system of planets is one          (e)B
    that has a

    a. companion star.      d. all of the above.
    b. high spin rate.      e. none of the above.
    c. mass about 4 times solar mass.


21. Some binary stars have dead companions.  Compared to the          (b)B
    mass of a live star, the mass of a dead companion is usually

    a. less.                c. the same.
    b. more.                d. ...depends on circumstances.


22. The Earth is accelerating towards the Sun.  If the Sun            (c)B
    collapsed into a black hole, this acceleration would

    a. increase.            c. stay the same.
    b. decrease.            d. stop.


23. The gravitational field at the surface of a star of radius       (c)B
    R has a certain field strength.  If the star collapses to
    become a black hole, the field strength at the same
    distance, R, will be

    a. greater.             c. no different.
    b. less.


24. If the Sun collapsed to a neutron star or a black hole, the      (d)B
    Earth would

    a. lose its oceans because of vaporization.
    b. spiral into the Sun's core.
    c. spiral away from the Sun's core.
    d. remain in its present orbit.
    e. undergo collpase also.


25. As more mass is captured in a black hole, the radius of the      (a)B
    event horizon

    a. increases.           c. stays the same.
    b. decreases.


26. As more mass is captured in a black hole, the radius of the      (b)B
    photon sphere

    a. increases.           c. remains the same.
    b. decreases.


27. Suppose one day we measure light from each galaxy in the universe (a)B
    to be blue-shifted.  This would mean that the universe is

    a. contracting.         d. all of the above.
    b. oscillating.         e. none of the above.
    c. expanding.


28. The three physical properties of a black hole are                (e)B

    a. mass, charge, temperature.
    b. temperature, density, charge.
    c. mass, angular momentum, density.
    d. mass, angular momentum, temperature.
    e. mass, angular momentum, charge.

29. Elements of atomic mass greater than iron are not made in     (c)C
    star cores because

       a. masses of heavier nuclei are less than the sum of
          their parts.
       b. radiation pressure at this stage overwhelms
          gravitational pressure.
       c. such reactions absorb rather than liberate energy.
       d. all of the above.
       e. none of the above.

30. The core temperature of a star must be hotter for helium      (c)C
    nuclei to fuse together than for hydrogen nuclei to fuse
    together because

       a. helium atoms have more mass than hydrogen atoms.
       b. helium is denser than hydrogen.
       c. helium has more electric charge than hydrogen.
       d. all of the above.
       e. none of the above.

# Appendix V

## Exponential Growth

Appendix 5: Exponential Growth

1. In an economy that has a steady inflation rate of 7 % per
   year, in how many years does a dollar loss half its value?        (b)A

       a. 7.            d. 20.
       b. 10.          e. none of these.
       c. 17.

2. If world population doubles in 40 years and world food
   production also doubles in 40 years, the number of people           (d)A
   starving in 40 years compared to now would be

       a. 1/4 as many.     d. twice as many.
       b. 1/2 as many.     e. four times as many.
       c/ the same.

3. Suppose a lily pond starts with a single lilypad, and
   doubles on each day until it is completely covered in              (c)A
   30 days.  On what day is the pond half covered?

       a. day 15.       d. none of these.
       b. day 20.       e. not enough information given.
       c. day 29.

4. Suppose that coal is taken from a mine at a steady rate with
   a constant  doubling time, and in the most recent doubling         (b)B
   time, 16384 carloads of coal were taken out.  How many
   carloads were taken out in all the previous doubling times?

       a. 8192 carloads.     d. 163840 carloads.
       b. 16383 carloads.    e. none of these.
       c. 32766 carloads.

5. Suppose a 7 % inflation rate holds steady for 70 years.  In
   70 years how much will a $1.00 cup of soup cost?                   (e)BM

       a. $2.         d. $100.
       b. $14.        e. $128.
       c. $64.

6. In the 1984 the population growth rate for the United
   States was 0.7 % and for Mexico it was 2.6 %.  At these            (b)BM
   rates about how long will it take the populations in each
   country to double?

       a. 10 years for the U.S., 2.7 years for Mexico.
       b. 100 years for the U.S., 27 years for Mexico.
       c. 1000 years for the U.S., 270 years for Mexico.
       d. not enough information is given.
       e. none of these.

7. Suppose the doubling time for bacteria growing in a jar      (d)BM
   is 1 minute, and the jar starts with one bacterium and fills
   to capacity in 60 minutes.  How many minutes passes before
   the jar fills to one quarter capacity?

      a. 15 min.           d. 58 min.
      b. 30 min.           e. none of these.
      c. 45 min.

8. If your wages are one dollar for the first day, two dollars   (d)CM
   for the second day, four dollars for the third day, and so
   on, how much will your combined wages be for 10 days?

      a. $20.             d. $1023.
      b. $511.          e. none of these.
      c. $512.